New edition of the #1 favorite!

MOORE and PERSAUD

The Developing Human

Clinically Oriented Embryology

FIFTH EDITION

HUMAN REPRODUCTIVE BIOLOGY

H U M A N
R E P R O D U C T I V E
B I O L O G Y

Second Edition

Sylvia S. Mader

McGraw
Hill

Boston   Burr Ridge, IL   Dubuque, IA   Madison, WI   New York   San Francisco   St. Louis
Bangkok   Bogotá   Caracas   Lisbon   London   Madrid
Mexico City   Milan   New Delhi   Seoul   Singapore   Sydney   Taipei   Toronto

McGraw-Hill Higher Education

A Division of The McGraw·Hill Companies

Book Team

Editor *Kevin Kane*
Developmental Editor *Carol Mills*
Production Editor *Diane Clemens*
Designer *Elise A. Burckhardt*
Art Editor *Miriam J. Hoffman*
Photo Editor *Robin Storm*
Permissions Editor *Gail Wheatley*
Visuals Processor *Andrêa Lopez-Meyer*
Developmental Visuals/Design Consultant *Marilyn A. Phelps*

President *G. Franklin Lewis*
Vice President, Publisher *George Wm. Bergquist*
Vice President, Operations and Production *Beverly Kolz*
National Sales Manager *Virginia S. Moffat*
Group Sales Manager *Vincent R. Di Blasi*
Vice President, Editor in Chief *Edward G. Jaffe*
Advertising Manager *Amy Schmitz*
Managing Editor, Production *Colleen A. Yonda*
Manager of Visuals and Design *Faye M. Schilling*
Production Editorial Manager *Julie A. Kennedy*
Production Editorial Manager *Ann Fuerste*
Publishing Services Manager *Karen J. Slaght*

President and Chief Executive Officer *Mark C. Falb*
Chairman of the Board *Wm. C. Brown*

Cover image by © Lennart Nilsson, *A Child Is Born,* Dell Publishing Co.
Copyediting by Laura Beaudoin

To my sister, Rhetta,
for her constant support

Brief Contents

Contents

II

Human Reproduction 69

Evolution, Behavior, and Population Concerns 161

Preface

This text presents human reproduction from the biological point of view. It is most appropriate for use by nonscience students who would like a biological understanding of human reproduction. The three parts of the book may be studied in whatever sequence the instructor desires.

Part I suggests that the purpose of reproduction is the passage of DNA, the genetic material, from one generation to the next. DNA makes up the genes that lie within the chromosome, and chromosomal inheritance is discussed prior to genetic inheritance. Sexual reproduction assures that the genotype of the offspring will differ from the genotype of either parent. The chances of an offspring receiving a certain combination of genes may sometimes be predicted, and this is particularly helpful when the offspring might receive a gene that causes a genetic disease. Both theoretical scientists and medical practitioners are presently emphasizing that many human illnesses are actually genetic diseases that can possibly be prevented. Techniques are available today and perhaps soon more will be available to control and to cure genetic diseases. The hope for cures lies in modern genetic research, including genetic engineering. The techniques of genetic engineering have led to a biotechnology industry, which is also discussed.

In part II, hormone action is explained before male and female reproductive anatomy, physiology, and sexual response are presented in depth. Following a description of fertilization, the major biological events of embryonic and fetal development are outlined. A discussion of birth-control procedures and devices emphasizes the effectiveness and the side effects of each method. Infertility and alternative methods of reproduction are also discussed.

Signs, symptoms, and cures of the most prevalent sexually transmitted diseases are reviewed, and the text also has a special supplement on AIDS.

In part III, the evolution of sexual reproduction is presumed to have increased biological fitness. That is, it is assumed that sexual reproduction has increased the likelihood of survival and chance of successful reproduction. Biologically fit organisms are adapted to their environment, and this adaptation includes anatomical, physiological, and behavioral adaptation. Sociobiology suggests that human reproductive behavior evolved to ensure the passage of genes.

The reproductive drive in humans has resulted in a very large human population. This population is undergoing exponential growth that could possibly outstrip the carrying capacity of the environment. The world is divided into more developed and less developed countries, and it is the less developed countries that contribute most to exponential growth. Human activities cause land, water, and air pollution, which threatens the integrity of the biosphere. However, a sustainable world is possible if population and economical growth is managed and accompanied by ecological preservation.

ACKNOWLEDGMENTS

Many people have contributed to this edition of *Human Reproductive Biology*. My editor, Kevin Kane, and developmental editor, Carol Mills, directed the efforts of all. Diane Clemens was the production editor, and Elise Burckhardt was the designer. They both worked with diligence and care. Kathleen Hagelston provided many new drawings that will be helpful to all users.

I especially want to thank the several instructors who reviewed the entire manuscript. They were

Francis M. Maxin *Community College of Allegheny*
Robert J. Caron *Bristol Community College*
Yvonne Patricia Boyd-Bartlett *York College of the City University of New York*
Terrill W. Tugel *Monroe Community College*

TO THE STUDENT

Human Reproductive Biology includes a number of aids that will help you study the topic successfully and enjoyably.

Part Introductions: An introduction to each part highlights the central ideas of that part and specifically tells you how the topics contribute to biological knowledge.

Boldfaced Terms: Terms that are pertinent to the topic being discussed appear in boldfaced print. They are defined in context, and all are defined in the glossary.

Tables and Illustrations: Several tables and illustrations appear in each chapter. The tables clarify complex ideas and summarize sections of the narrative. The photographs and drawings have been carefully chosen and designed to help you visualize structures and processes.

Boxed Readings: Several boxed readings expand, in an interesting way, on the core information presented in each chapter. They ask you to consider and evaluate a matter of current social concern.

Human Issue Boxes: Human Issue boxes are included in every chapter throughout the text. These general discussion boxes will stimulate your interest and thought, especially about how the chapter topics can be applied to human personal concerns.

Chapter Summaries: Chapter summaries offer a concise review of material. You may read them before beginning the chapter to preview the topics of importance, and you may also use them to refresh your memory after you have a firm grasp of the concepts presented.

Key Terms Lists: The key terms list, which appears at the end of each chapter, contains all the boldfaced terms used in the chapter. Each term is accompanied by a page number indicating where it is introduced and defined.

Review Questions: The review questions allow you to test your understanding of the information in the chapter. When you can successfully answer each question, you have mastered the content of the chapter. Answers to the genetics questions in chapters 2 and 3 appear in Appendix B.

Critical Thinking Questions: All persons need to learn to think critically. The critical thinking questions ask you to apply your recently acquired knowledge to new and different situations. Suggested answers appear in the *Instructor's Manual*.

Further Readings: If you would like more information about a particular topic or are seeking references for a research paper, a listing of articles and books to help you get started appears at the end of each part. Usually the entries are *Scientific American* articles and books that expand on the topics covered in the chapters.

AIDS Supplement: Because of the current concern and recent discoveries in AIDS research, a special section has been set aside for extensive treatment of this disease. This section includes the origin, transmission, stages, treatment, prevention, and future of the disease. The AIDS supplement follows chapter 11.

Appendixes: Appendix A lists the acronyms used in the text, along with the complete terms.

Glossary and Index: An end-of-book glossary and index are included. The glossary allows you to review the definitions of the boldfaced terms that appear in the text. The index helps you to locate topics quickly.

ADDITIONAL AIDS

Instructor's Manual/Test Item File: The Instructor's Manual/Test Item File, prepared by Arthur Cohen of Massachusetts Bay Community College, is designed to assist instructors as they plan and prepare for courses using *Human Reproductive Biology*. An outline and general discussion are provided for each chapter; together these give the overall rationale for the chapter. Suggested answers to the critical thinking questions are found in this manual. A list of suggested films for each chapter and a list of film suppliers are also provided.

The test item file contains many objective test questions and several essay questions for each chapter. As noted, the objective questions require different levels of thinking.

Transparencies: This edition is accompanied by 25 transparencies in two and four colors. The transparencies feature text illustrations with oversized labels, facilitating their use in large lecture rooms. They are available free to all adopters.

TestPak: wcb TestPak is a computerized system that enables you to make up customized exams quickly and easily. Test questions can be found in the Test Item File, which is printed in your instructor's manual or as a separate packet. For each exam you may select up to 250 questions from the file and either print the test yourself or have wcb print it.

HUMAN REPRODUCTIVE BIOLOGY

Down syndrome child participating in Special Olympics. Down syndrome is an inherited condition.

I

Human Inheritance

Sexual reproduction in humans requires sex cell formation, fertilization, and development. Each sex cell carries half of the total number of chromosomes as the body cells. This helps ensure that no child (except identical twins) receives exactly the same combination of chromosomes. The genes are on the chromosomes. It is sometimes possible to determine the chances of an offspring receiving a particular chromosome and gene. Therefore, if the genetic makeup of the parent is known, it may be possible to determine the chances of a child inheriting a genetic disease.

Genes are now known to be constructed of DNA. They control not only the characteristics of the individual but also the cell. DNA contains a code for the sequence of amino acids in proteins, which are synthesized in the cell. This is how DNA controls the makeup and functioning of the cell. In the past several years, knowledge of DNA structure and function has led to a new industry—biotechnology—which has already contributed advances in the fields of agriculture and medicine. One day it may be possible to cure human genetic diseases. ■

1

Chromosomes and Chromosomal Inheritance

Every human being is made up of billions of cells. Figure 1.1 shows a generalized cell, which can be divided for convenience into two parts: the cytoplasm and the nucleus. The **cytoplasm,** sometimes defined as the liquid portion of the cell, surrounds the central **nucleus** and contains various small bodies called organelles. Each organelle has a specific support function for the cell, and those of interest to us will be discussed in chapter 4. Inside the nucleus are the **chromosomes,** which are filamentous structures that contain a variety of genes. **Genes** determine what the cell is like and what the individual is like. In a nondividing cell the chromosomes are indistinct and diffuse, but in a dividing cell the chromosomes become short and thick and may be counted using a high-powered microscope. Normal humans contain 23 pairs of chromosomes, or 46 chromosomes altogether, in each nucleus of somatic cells. *Somatic cells* are body cells such as skin, muscle, nerve, and liver cells.

A cell may be photographed under a high-powered microscope just prior to division to obtain a picture of the chromosomes. The chromosomes may be cut out of an enlarged picture and arranged by pairs (fig. 1.2). Pairs of chromosomes are recognized by the fact that they are of

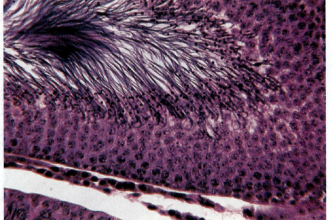

Cross section of seminiferous tubule where sperm production is occurring.

the same length and that the centromere, a constriction, is similarly placed. They may also be stained, in which case the pairs have the same banding patterns.

The resulting display of pairs of chromosomes is called a **karyotype.** Although both males and females have 23 pairs of chromosomes, one of these pairs is of unequal length in the male. The larger chromosome of this pair is called the X and the smaller is called the Y. Females have two X chromosomes in their karyotype. The X and Y chromosomes are called the **sex chromosomes** because they contain the genes that determine sexual characteristics of males and females.[1] The other chromosomes, known as **autosomes,** include all the pairs of chromosomes except the X and Y chromosomes. Notice that each pair of autosomes in the human karyotype is numbered from 1 to 22.

As figure 1.2 shows, prior to division each chromosome is composed of two identical parts, called **chromatids,** which are held together at a centromere.

1. Homosexuality is a behavioral pattern unrelated to the sex chromosomes; homosexual men have XY chromosomes, and lesbian women have XX chromosomes.

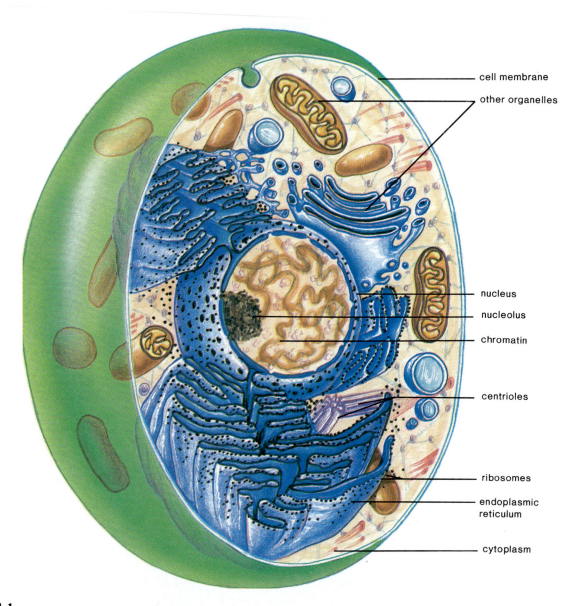

cell membrane

other organelles

nucleus

nucleolus

chromatin

centrioles

ribosomes

endoplasmic
reticulum

cytoplasm

Figure 1.1

The human cell has many small bodies called organelles. The nucleus is a central organelle that contains a nucleolus and chromatin, a substance that becomes chromosomes when the cell divides. The chromosomes contain the genes and determine what the cell is like and what the individual is like. The rough endoplasmic reticulum is a system of tubules and saccules that are studded with ribosomes. We will be discussing these parts of the cell in a later chapter. The cytoplasm, the liquid part of the cell, is contained by the cell membrane.

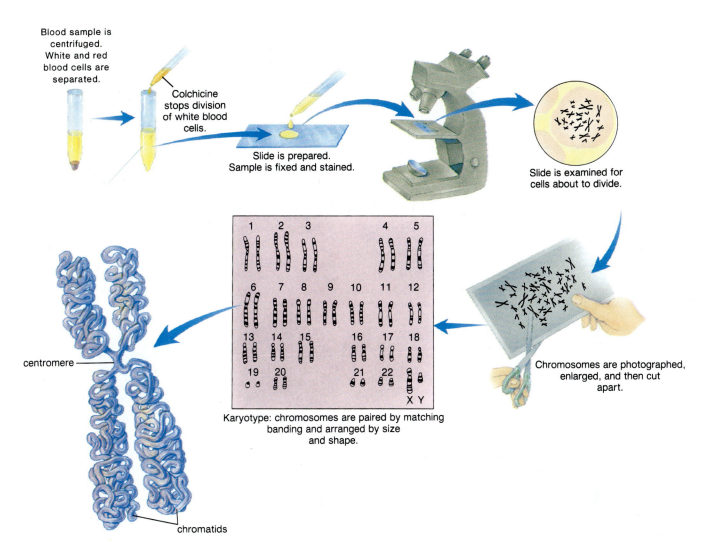

Blood sample is centrifuged. White and red blood cells are separated.

Colchicine stops division of white blood cells.

Slide is prepared. Sample is fixed and stained.

Slide is examined for cells about to divide.

Chromosomes are photographed, enlarged, and then cut apart.

Karyotype: chromosomes are paired by matching banding and arranged by size and shape.

centromere

chromatids

Figure 1.2

A human karyotype is prepared by the steps illustrated here. A karyotype is a display of a person's chromosomes (see box); every person has 22 pairs of autosomes and one pair of sex chromosomes. This is a karyotype of a male because it has an X and Y chromosome; females have two X chromosomes. The enlargement of a chromosome (*far left*) shows that the chromosomes in a karyotype consist of two chromatids held together by a centromere. This is the appearance of chromosomes just before they divide.

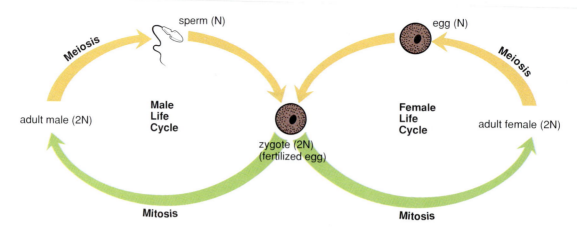

Figure 1.3

Life cycle of animals. Notice that there are two types of cell division—mitosis and meiosis. For the male life cycle, after the sperm fertilizes the egg, the diploid zygote divides mitotically as the individual becomes the adult male. Then, meiosis occurs during the production of haploid sperm. For the female life cycle, after the sperm fertilizes the egg, the zygote divides mitotically as the individual becomes the adult female. Then, meiosis occurs during the production of haploid eggs.

HUMAN LIFE CYCLE

The human life cycle requires **sexual reproduction,** in which the **sperm** of the male fertilizes the **egg** in the female; the resulting **zygote** develops over the next 9 months into the newborn infant, who grows to be an adult (fig. 1.3).

Two types of cell division occur during the human life cycle. **Meiosis** occurs as a part of **gametogenesis,** which is the production of gametes. This term is used collectively to mean the production of sex cells, that is, the sperm and egg. Because of meiosis, a sperm contains only 23 chromosomes and an egg contains only 23 chromosomes. The zygote contains 23 *pairs* of chromosomes; one of each pair is contributed by the father's sperm, and one of each pair is contributed by the mother's egg.

The second type of cell division, called **mitosis,** occurs whenever growth takes place, such as when the zygote develops into a newborn. The zygote receives 46 chromosomes, and as a result of mitosis every cell in the body thereafter has 46 chromosomes. Each cell in the body has a copy of the gene-containing chromosomes originally contributed by the parents' gametes. Offspring tend to resemble their parents because the parents and child share common chromosomes and genes.

CELL DIVISION

Mitosis

Mitosis is cell division in which the chromosome number stays constant; that is, the original cell, called the *mother* cell, has 46 chromosomes, and the two *daughter,* or re-

sulting, cells also have 46 chromosomes. When a cell has the full number of possible chromosomes, it is said to have the **diploid,** or 2N, number of chromosomes. Mitosis, then, is cell division in which 2N $\longrightarrow$ 2N.

Figure 1.4 shows a diagram of a mother cell that for simplicity's sake has only two pairs of chromosomes because all the other chromosomes behave similarly during mitosis. Notice that each chromosome goes through a cycle in which it is at first a singled chromosome, then a duplicated chromosome, and then a singled chromosome again.

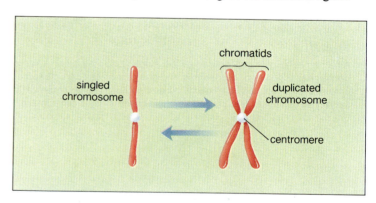

Just before cell division, the chromosomes duplicate. This involves the replication of DNA, the genetic material, and will be discussed in chapter 4. During mitosis, the chromatids of a duplicated chromosome separate, and one chromatid from each duplicated chromosome goes to each daughter cell. *After chromatids separate, they are called chromosomes.* Therefore, each daughter cell receives the same number and kinds of chromosomes as the mother cell.

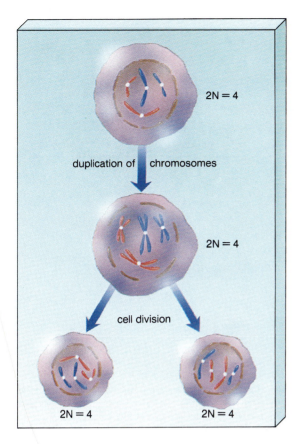

Figure 1.4

Mitosis overview. Following duplication, each chromosome in the mother cell contains chromatids. During mitotic division, the chromatids separate so that daughter cells have the same number and kinds of chromosomes as the mother cell. (The blue chromosomes in the mother cell were inherited from one parent, and red chromosomes were inherited from the other parent.)

Mitosis is a part of the cell cycle, which also includes interphase (fig. 1.5). During interphase the cell resembles the cell in figure 1.1 and is carrying on its normal activities. When and if the cell is going to divide, chromosome duplication occurs; then each chromosome is composed of two chromatids held together by a centromere. Following mitosis, each chromosome is single again. Specialized cells such as muscle and nerve cells "break out" of the cell cycle and no longer divide.

Figure 1.6 diagrams the process of mitosis. The process actually requires several stages, during which the nuclear envelope disappears and a spindle apparatus with spindle fibers forms. The chromosomes are attached to the spindle fibers by their centromeres, and the chromosomes move to the center of the mother cell before each centromere splits and the chromatids separate. The separated

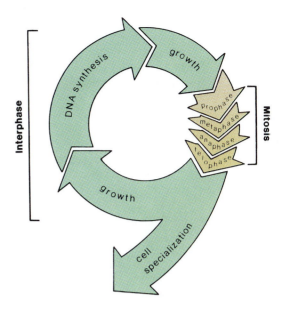

Figure 1.5

The cell cycle. Immature cells go through a cycle that consists of mitosis and interphase. During interphase, there is growth before and after chromosome duplication. Eventually, some daughter cells "break out" of the cell cycle and become specialized cells performing a specific function.

chromatids, now called chromosomes, move away from each other into the newly forming daughter cells.

Figure 1.7 shows cells undergoing mitosis. Mitosis is the type of cell division required for growth and repair of body cells. The process of mitosis assures that each cell in the body has the same number and kinds of chromosomes and genes.

The term *cloning* is sometimes used to refer to the asexual production of individuals from mature cells. As you would suspect, only embryonic animal cells seem to retain the capability of being able to divide and begin development again. An adult cell is too specialized to begin development. Research is still going on to overcome this difficulty so that a human could be cloned from an adult cell.

·   HUMAN ISSUE   ·

If one day it is possible for humans to make clones of themselves, should each and every person be allowed to do so? Or should only those persons who are judged to have the best of personal attributes be allowed to clone themselves? Who would make judgments about the worthiness of individuals to be cloned?

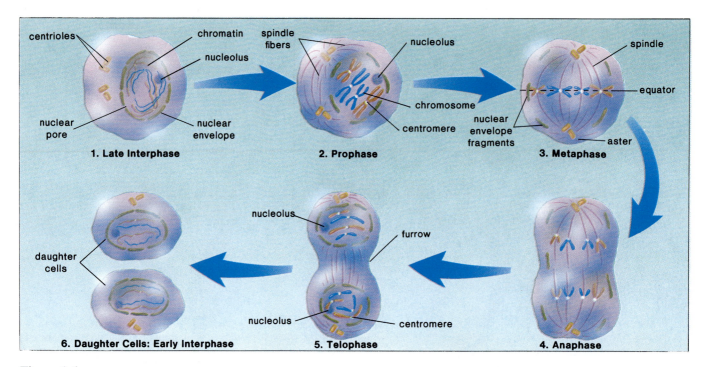

Figure 1.6

Mitotis in detail. Mitosis has four stages, excluding interphase and daughter cells. (1) During interphase, the nuclear envelope is intact and a small body, the nucleolus, is clearly visible in the nucleus. The chromosomes have duplicated, as have the centrioles; they are organelles associated with the formation of the spindle. (2) During prophase the nuclear envelope and nucleolus are disappearing, and the spindle fibers are appearing as the centrioles move apart. (3) During metaphase the chromosomes are lined up at the middle of the spindle. (4) During anaphase the chromatids separate and the resulting daughter chromosomes move apart. (5) During telophase, the nuclear envelope and nucleolus are reappearing, the spindle fibers are disappearing and a furrow (indentation) appears. (6) Furrowing of the cell membrane divides the cytoplasm between two daughter cells.

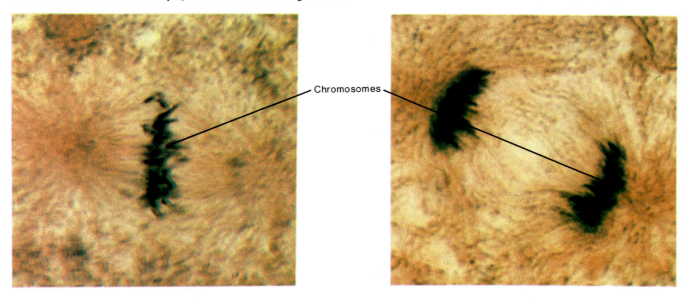

Figure 1.7

Photomicrographs of cells undergoing mitosis. *a.* This is metaphase—the chromosomes are at the equator of the spindle. *b.* This is late anaphase—the chromosomes are at the poles of the spindle but furrowing which divides the cytoplasm has not begun.

Meiosis

Meiosis is cell division in which the chromosome number is halved. In humans, the mother cell has 46 chromosomes, but at the completion of meiosis the daughter cells have just 23 chromosomes, which are not paired. Half the chromosome number is called the **haploid,** or N, number of chromosomes. Meiosis is cell division in which 2N ⟶ N.

Meiosis requires two cell divisions, called meiosis I and meiosis II (fig. 1.8). As a result there are four daughter cells, each having the haploid number of chromosomes. Before meiosis begins, duplication of chromosomes has occurred. During meiosis I, similar duplicated chromosomes in the mother cell come together in pairs. Then, the duplicated chromosomes of each pair separate so that the daughter cells receive only one from each chromosomal pair. Notice that no duplication of chromosomes need occur before the second meiotic division because each chromosome is already duplicated. During meiosis II, the chromatids of each duplicated chromosome separate. After the chromatids separate, they are called chromosomes. Therefore, each of four daughter cells has half the number of chromosomes as the mother cell and one of each kind of chromosome as compared to the original mother cell.

Figure 1.9 shows the process of meiosis in detail. The process actually requires several stages. Each series of stages occurs twice—once during meiosis I and again during meiosis II. Whereas mitosis is the cell division for growth and repair, meiosis occurs only during the production of gametes. The production of sperm cells in males is called **spermatogenesis;** production of the egg cell in females is called **oogenesis.**

Spermatogenesis

The left-hand side of figure 1.10 shows meiosis as it occurs in the testes of males during spermatogenesis, or sperm production. In the testes new primary spermatocytes are produce continually. Primary spermatocytes with 46 chromosomes undergo meiosis I to produce daughter cells that have 23 chromosomes. These secondary spermatocytes undergo meiosis II to produce four daughter cells (spermatids) that contain 23 singled chromosomes. Two out of every four sperm ordinarily carry an X chromosome and two carry a Y chromosome.

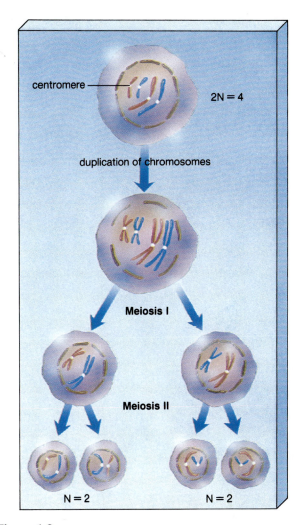

Figure 1.8

Overview of meiosis, the type of cell division that produces the sex cells; sperm in males and eggs in females. Duplication of chromosomes occurs before meiosis begins. The diploid (2N) number of chromosomes in this mother cell is 4. During meiosis, the chromosomes of each pair separate so that the daughter cells have one of each kind of duplicated chromosome. During meiosis II, chromatids separate so that the four resulting daughter cells have one-half the number and one of each kind of chromosome as the original mother cell. The haploid (N) number of chromosomes in the daughter cells is 2.

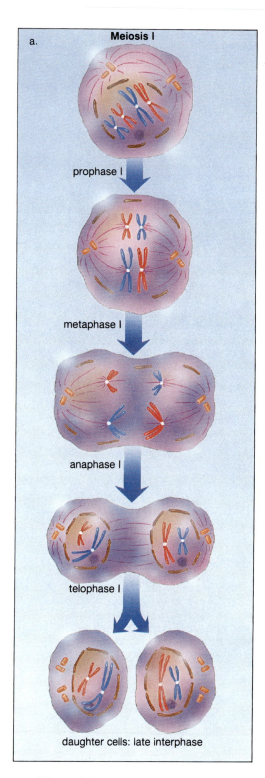

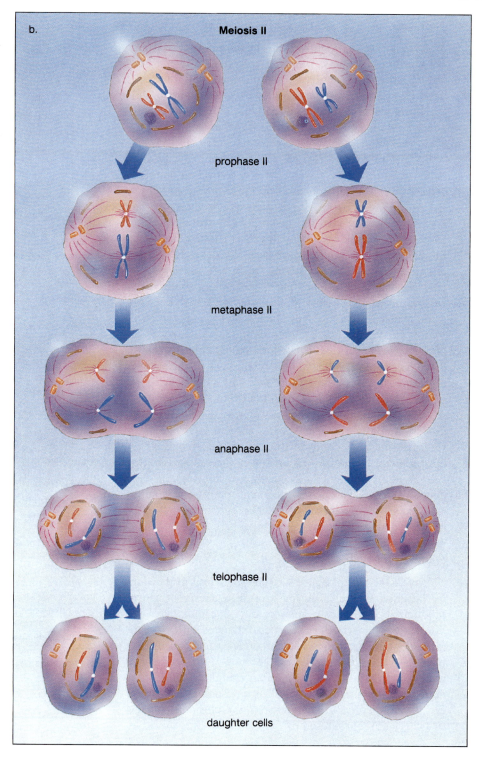

Figure 1.9

a. Meiosis I. During *prophase I*, the spindle appears as the nuclear envelope and nucleolus disappear. At *metaphase I*, the pairs of chromosomes line up at the middle of the spindle. During *anaphase I*, the chromosomes of each pair separate and move apart. During *telophase I*, the nuclear envelope and nucleolus reappear as the spindle disappears. The cell membrane furrows to give two complete cells. Daughter cells have one from each pair of chromosomes. *b.* Meiosis II. During *prophase II*, a spindle appears as the nuclear envelope and nucleolus disappear. During *metaphase II*, the chromosomes line up at the middle of the spindle. During *anaphase II*, chromatids separate and move apart. During *telophase II*, the spindle disappears as the nuclear envelope reappears. The cell membrane furrows to give daughter cells each with the haploid number of chromosomes.

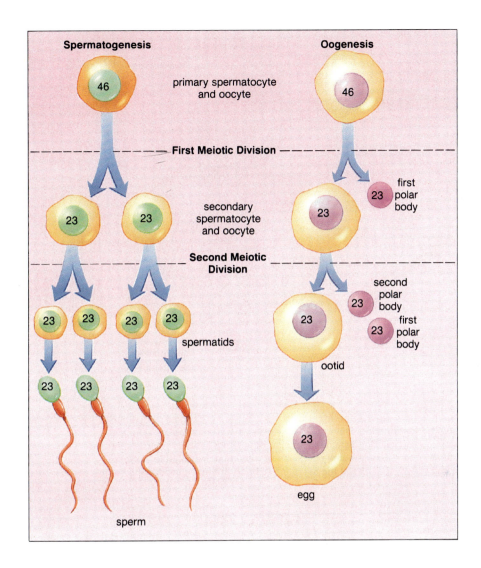

Figure 1.10

Spermatogenesis and oogenesis in humans. In the testes, sex organs of males, primary spermatocytes with 46 chromosomes undergo meiosis I and the resulting cells have 23 (duplicated) chromosomes each. These cells undergo meiosis II and there are four final cells each with 23 (singled) chromosomes. In the ovaries, the sex organs of females, primary oocytes with 46 chromosomes undergo meiosis I and the resulting cells have 23 (duplicated) chromosomes each. One of these cells is a polar body that will eventually disintegrate. Meiosis II produces a large cell that will become the egg and another polar body, each with 23 (singled) chromosomes. When a sperm fertilizes an egg, the new individual has 46 chromosomes again.

Oogenesis

The right-hand side of figure 1.10 shows meiosis as it occurs in the ovaries of females during oogenesis, or egg production. A female is born with all the primary oocytes she will ever have. Oogenesis also differs from spermatogenesis in that meiosis I results in one viable daughter cell (secondary oocyte) and one **polar body.** A polar body is a much smaller cell that is nonfunctional. Meiosis II also results in one daughter cell and another polar body. Notice that only the daughter cell, containing most of the cytoplasm, matures into an egg (fig. 1.11), and therefore fe-

males produce only one egg per oogenesis. Since the primary oocyte contained only X chromosomes, all eggs ordinarily contain one X chromosome.

There are other differences between spermatogenesis and oogenesis. Spermatogenesis, once started, always goes to completion and mature sperm result. In contrast, oogenesis does not necessarily go to completion. Only if a sperm fertilizes the maturing egg does that egg undergo meiosis II; otherwise it simply disintegrates. Regardless of this complication, however, both the sperm and egg contribute the haploid number of chromosomes to the new individual.

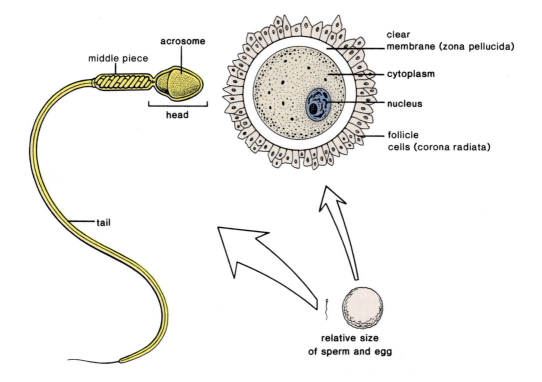

Figure 1.11

The egg is spherical and much larger than the sperm because it contains a great deal of cytoplasm. The outer membrane of the egg is convoluted and often surrounded by adhering cells collectively called the corona radiata. The mature sperm, which measures 1/500 inch, consists of a head, middle piece, and tail. The flattened and almond-shaped head contains the nucleus, very little cytoplasm, and a caplike acrosome at the anterior end. The acrosome contains enzymes that are capable of digesting away the outer covering of the egg. The middle piece produces the energy that allows the tail to move back and forth, accounting for the ability of the sperm to swim.

● **HUMAN ISSUE** ●

In a recent poll of Americans, about two-thirds of those interviewed said they would like to know the sex of their child before birth, and about one-fourth thought it is justifiable to treat the father's sperm in a medical laboratory to increase the odds of conceiving a boy or a girl. Also, 1 person out of 20 thought abortions are justified if the child is not of the desired gender.

Should the sex of a child be of major importance? For example, is a man more manly if he has a son? Should couples simply take their chances as to the sex of their child, or is it better to predetermine the sex? What methods of controlling the sex of children do you think should be banned?

Egg production occurs once a month in females, while sperm production occurs continuously in males. At the time of ejaculation during intercourse, males emit as many as 400 million or more sperm (chapter 7).

CHROMOSOMAL INHERITANCE

As mentioned, an individual normally receives 22 autosomal chromosomes and 1 sex chromosome from each parent at fertilization. The sex of the newborn child is determined by whether a Y- or X-bearing sperm fertilizes the egg. While it is obvious that there is a 50% chance, all other factors being equal, of having a girl and a 50% chance of having a boy, it is possible to illustrate this probability with a **Punnett Square** (fig. 1.12).

In the square, all possible sperm are lined up on one side and all possible eggs are lined up on the other side (or vice versa); then every possible mating is considered. When this is done with regard to sex chromosomes, the results show one female to each male. However, for reasons that are not clear, more males than females are miscarried, and more females are born than males. After birth, this trend continues until there is a dramatic reversal of the ratio of males to females among seniors (table 1.1).

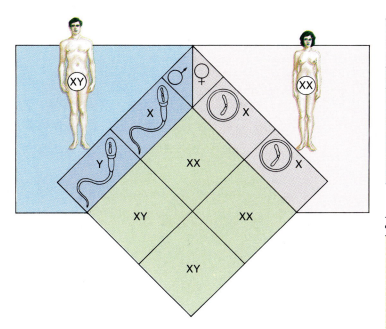

Figure 1.12

In a Punnett Square, all possible sperm fertilize all possible eggs so that the chances of having a particular offspring can be determined. Here sperm and eggs are shown as carrying only a sex chromosome; actually, of course, they also carry 22 autosomes. The offspring (located within the squares) are either male or female depending on whether they received an X or Y from the male parent.

Abnormal Autosome Inheritance

Sometimes individuals are born with either one or two too many or too few chromosomes. It is possible also that even though there is the correct number of chromosomes, one chromosome may be defective in some way.

The most common autosomal abnormality is seen in individuals who have **Down syndrome** (fig. 1.13). A **syndrome** is a group of symptoms that always occur together in an individual due to the presence of an abnormal condition. Down syndrome is easily recognized. Its characteristics include a short stature; an oriental-like fold of the eyelids; stubby fingers; a wide gap between the first and second toes; a large, fissured tongue; a round head; and

TABLE 1.1	SEX RATIOS IN THE UNITED STATES	
Age	**Males:Females**	
Birth	106:100	
18 years	100:100	
50 years	85:100	
85 years	50:100	
100 years	20:100	

mental retardation that can sometimes be severe. Also seen is an increased tendency toward leukemia, cataracts, and Alzheimer's disease, a form of senility.

It is now known that the genes causing Down syndrome are located on the lower third of chromosome 21, and extensive research has led to the discovery of which specific genes are responsible for the characteristics of the syndrome. For example, researchers have located the gene, called the Gart gene, believed to be responsible for mental retardation. It is hoped that someday a way to control the expression of this gene before birth will be found, so that this particular symptom of Down syndrome will not appear.

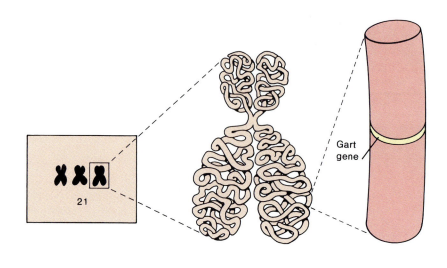

a.

b.

Figure 1.13

Down syndrome is an inherited condition. *a.* Common characteristics include a wide, rounded face and a fold of the upper eyelids. Mental retardation, along with an enlarged tongue, makes it difficult for persons with Down syndrome to learn to speak coherently. *b.* Karyotype of an individual with Down syndrome has an extra number-21 chromosome. More sophisticated technologies allow investigators to pinpoint the location of specific genes associated with the syndrome. The Gart gene may account for the mental retardation seen in persons with Down syndrome.

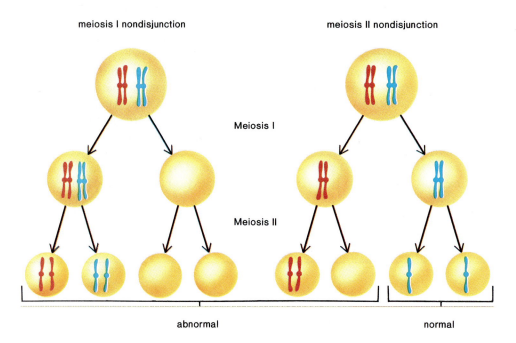

Figure 1.14

Nondisjunction during oogenesis. Nondisjunction can occur during meiosis I if the chromosomes of a pair fail to separate and during meiosis II if the chromatids fail to separate completely. In either case, some of the abnormal eggs carry an extra chromosome. Nondisjunction of the number-21 chromosome leads to Down syndrome.

TABLE 1.2 FREQUENCY AND EFFECTS OF THE MOST COMMON CHROMOSOME ABNORMALITIES IN HUMANS

Syndrome	Sex	Chromosomes	Frequency Miscarriages	Births	Fertility
Down					
General	M or F	Extra 21	1/40	1/700	Infertile
Mothers over 40				1/70	
Turner	F	XO	1/18	1/5,000	Sterile
Metafemale	F	XXX	0	1/700	Infertile
Klinefelter	M	XXY	0	1/2,000	Sterile
XYY	M	XYY	?	1/2,000	—

Nondisjunction

Persons with Down syndrome usually have three number-21 chromosomes because the egg had two number-21 chromosomes instead of one. In about a quarter of individuals, the sperm had the extra chromosome. Either the chromosomes of a pair or the chromatids failed to separate completely and instead went into the same daughter cell (fig. 1.14). Either of these occurrences is called **nondisjunction.**

It would appear that nondisjunction is most apt to occur in the older female, whose oocytes are as old as she is (chapter 8), since children with Down syndrome are usually born to women over age 40 (table 1.2). If the older woman wishes to know whether or not her unborn child is affected by Down syndrome, she may elect to undergo **chorionic villi testing** or **amniocentesis,** two procedures discussed in the chapter 9 reading, "Detecting Birth Defects." Following either procedure, a karyotype can reveal whether the child has Down syndrome. As also discussed in the reading, a new procedure available only to those undergoing in vitro fertilization can be used to make sure that a normal egg is being fertilized.

Deletion

Another chromosomal abnormality called deletion is responsible for a syndrome known as **Cri du Chat.** The condition derives its name from the French, and affected infants meow like a kitten when they cry. More important, perhaps, is the fact that they tend to have a small head with malformations of the face and body (fig. 1.15). Mental defectiveness usually causes retarded development. Chromosomal analysis shows that a portion of chromosome number 5 is missing (deleted), while the other number-5 chromosome is normal.

Abnormal Sex Chromosome Inheritance

Individuals are sometimes born with the sex chromosomes XO (Turner syndrome), XXX (metafemale), XXY (Klinefelter syndrome), and XYY (table 1.2). Individuals with a Y chromosome are always male no matter how many X chromosomes there may be; however, at least one X chromosome is needed for survival. An embryo with only one or more Y chromosomes dies spontaneously long before delivery.

Abnormal sex chromosome constituencies may result from nondisjunction of the sex chromosomes during oogenesis or spermatogenesis (fig. 1.16). Nondisjunction during either meiosis I or meiosis II of oogenesis can lead to an egg with two X chromosomes or no X chromosomes. Nondisjunction during meiosis I of spermatogenesis can lead to a sperm that has no sex chromosome or both an X and a Y chromosome. Nondisjunction during meiosis II of spermatogenesis can lead to a sperm with two X chromosomes or two Y chromosomes. In other words the chromatids of either an X or Y are separate, but they have both gone into the same gamete. Fertilization of abnormal gametes with normal gametes results in individuals with the syndromes noted in figure 1.16.

An XO individual with **Turner syndrome** has only one sex chromosome, an X; the O signifies the absence of the second sex chromosome. Turner females are short, have a broad chest, and may have congenital heart defects. Because the ovaries never become functional these females do not undergo puberty or menstruate, and there is a lack of breast development. Although no overt mental retardation is reported, Turner females show reduced skills in interpreting spatial relationships.

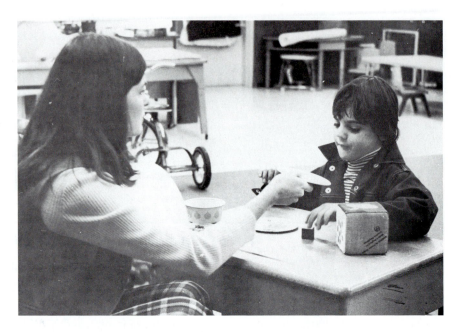

Figure 1.15

Child with Cri du Chat syndrome receiving educational therapy. Infants with this syndrome meow like a kitten when they cry. They also have malformations of the head, face, and body all due to a missing portion of chromosome number 5. Chromosome number 5 is an autosomal chromosome.

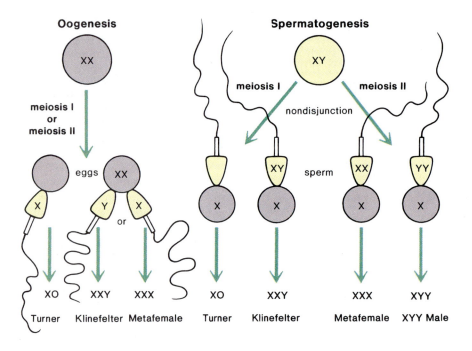

Figure 1.16

Left, nondisjunction of sex chromosomes during oogenesis followed by fertilization with normal sperm results in the conditions noted. Right, nondisjunction of sex chromosomes during spermatogenesis followed by fertilization of normal eggs results in the conditions noted. Turner syndrome (XO) individuals are females who never mature properly; Klinefelter syndrome (XXY) are males who show breast development; Metafemales (XXX) are females who may have reproductive problems; XYY males tend to be tall with barely normal intelligence.

When an egg with two X chromosomes is fertilized by an X-bearing sperm, a **metafemale** with three X chromosomes results. It might be supposed that the XXX female with 47 chromosomes would be especially feminine, but this is not the case. Although in some individuals there is a tendency toward learning disabilities, most metafemales have no apparent physical abnormalities except they may have menstrual irregularities, including early onset of menopause.

When an egg with two X chromosomes is fertilized by a Y-bearing sperm, a male with **Klinefelter (XXY) syndrome** results. This individual is sterile, the testes are underdeveloped, and there may be some breast development. These phenotypic abnormalities are not apparent until puberty, although some evidence of subnormal intelligence may be apparent before this time.

When a sperm with two Y chromosomes fertilizes an X-bearing egg, an **XYY male** results. Such males are usually taller than average, suffer from persistent acne, and tend to have barely normal intelligence. At one time it was suggested that these men were likely to be criminally aggressive, but it has since been shown that the incidence of such behavior is no greater than that among normal XY males.

Notice that there is no YO syndrome. As mentioned, embryos that contain only a single Y chromosome die because at least one X is required for survivial. ∎

SUMMARY

Human beings are made up of cells, each containing a nucleus with 46 chromosomes. A karyotype is a pictorial display showing the chromosomes arranged in 22 pairs of autosomal chromosomes and 1 pair of sex chromosomes. An individual with the sex chromosomes XX is a female and one with the sex chromosome XY is a male.

The human life cycle depends on sexual reproduction and two types of cell division—mitosis, in which the chromosome number stays constant ($2N \longrightarrow 2N$), and meiosis, in which

the chromosome number is halved ($2N \longrightarrow N$). Mitosis occurs during growth and repair of cells and results in two daughter cells each with 46 daughter chromosomes. Meiosis occurs during gametogenesis, the production of sex cells, and results in four daughter cells each with 23 chromosomes, one of each kind. In males, meiosis is a part of spermatogenesis and in females it is a part of oogenesis. When the sperm fertilizes the egg during sexual reproduction, the resulting zygote has 46 chromosomes, or 23 pairs. One

chromosome of each pair was derived from the sperm, and one was derived from the egg.

The major inherited autosomal abnormality is Down syndrome, in which the individual inherits three number-21 chromosomes due to nondisjunction during gamete formation. Cri du Chat, which is due to a deletion, also occurs. Examples of abnormal sex chromosome inheritance due to nondisjunction are Turner syndrome (XO), Klinefelter syndrome (XXY), XYY males, and metafemales (XXX).

KEY TERMS

REVIEW QUESTIONS

1. Draw a diagram to describe the human life cycle. Include mitosis and meiosis and either spermatogenesis or oogenesis. Use the appropriate number to denote the chromosome number of all structures.

2. Draw a generalized diagram for mitosis. (a) In each cell, put the notation 2N or N as appropriate. (b) Sketch an autosomal pair of chromosomes in the mother cell, and show what happens to the chromosomes during the process of cell division.

3. Draw a generalized diagram for meiosis. (a) In each cell, put the notation 2N or N as appropriate. (b) Place an autosomal pair of chromosomes in the mother cell, and show what happens to the chromosomes during the process of meiosis.

4. List several differences between mitosis and meiosis considering the following: (a) the purpose, (b) the occurrence in the body, (c) the number of divisions, (d) the number of daughter cells, (e) the number changes of the chromosomes, and (f) the resulting number of chromosomes in the daughter cells.

5. Diagram spermatogenesis and oogenesis. Note four differences between the two processes.

6. Name two inherited syndromes caused by autosomal chromosome abnormalities. State the abnormality in each and describe the appearance of the affected individual.

7. Why would a woman undergo chorionic villi testing and amniocentesis? How can a karyotype indicate that a baby will be abnormal at birth?

8. Name two syndromes involving sex chromosome abnormalities in females. What are the sex chromosomes for each? Describe the appearance of the individual for each.

9. Name two syndromes involving sex chromosome abnormalities in males. What are the sex chromosomes for each? Describe the appearance of the individual for each.

10. What is nondisjunction? Draw diagrams for autosomal nondisjunction during meiosis I and meiosis II. Draw another diagram for sex chromosome nondisjunction during spermatogenesis and oogenesis.

CRITICAL THINKING QUESTIONS

1. Assume a mother cell with four chromosomes. Which of the following two cells is a daughter cell following mitosis? a daughter cell following meiosis? How do you know? In what two ways do the chromosomes of the cells differ?

2. Chromosomes contain the genes that occur as alleles. For example, *A* is the allele of *a* and vice versa. One member of a pair of chromosomes contains allele *A* and the other member contains allele *a*. Can both alleles be in the same daughter cell following meiosis, or can only one allele, either *A* or *a,* be in the same daughter cell? How do you know?

3. Assume that one chromosome contains two different alleles, *A* and *B*. Will both *A* and *B* alleles be in one daughter cell following meiosis? How do you know?

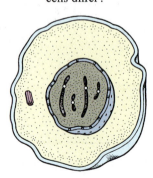 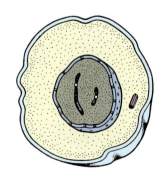

CHAPTER

2

Genes and Autosomal Inheritance

Genes, the units of heredity that control specific characteristics of an individual, are arranged in a linear fashion along the chromosomes. In this chapter, we will look at the inheritance of genes that are on the autosomes, the nonsex chromosomes.

GENES

Alternate forms of a gene having the same position (locus) on a pair of chromosomes and affecting the same trait are called **alleles.** It is customary to designate an allele by a letter in terms of the specific characteristic it controls; a **dominant allele** is assigned a capital letter, while a **recessive allele** is given the same letter lowercased. In humans, for example, unattached (free) earlobes are dominant over attached earlobes, so a suitable key would be *E* for unattached earlobes and *e* for attached earlobes.

Since autosomal alleles occur in pairs, the individual normally has two alleles for a characteristic. Just as one of each pair of chromosomes is inherited from each parent, so too is one of each pair of alleles inherited from each parent.

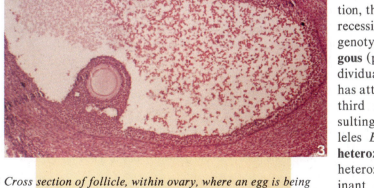

Cross section of follicle, within ovary, where an egg is being produced.

Figure 2.1 shows three possible fertilizations and the resulting genetic makeup of the zygote and, therefore, the individual. In the first instance, the chromosome of both the sperm and egg carries an *E*. Consequently, the zygote and subsequent individual have the alleles *EE,* which may be called a **homozygous** (pure) **dominant** genotype (table 2.1). The word **genotype** refers to the genes of the individual. A person with genotype *EE* obviously has unattached earlobes. The physical appearance of the individual, in this case unattached earlobes, is called the **phenotype.**

In the second fertilization, the zygote received two recessive alleles (*ee*), and the genotype is called **homozygous** (pure) **recessive.** An individual with this genotype has attached earlobes. In the third fertilization, the resulting individual has the alleles *Ee,* which is called a **heterozygous** genotype. A heterozygote shows the dominant characteristic; therefore, the phenotype of this individual is unattached earlobes.

These examples show that a dominant allele contributed from only one parent can bring about a particular phenotype. A recessive allele needs to be given from both parents to bring about the recessive phenotype.

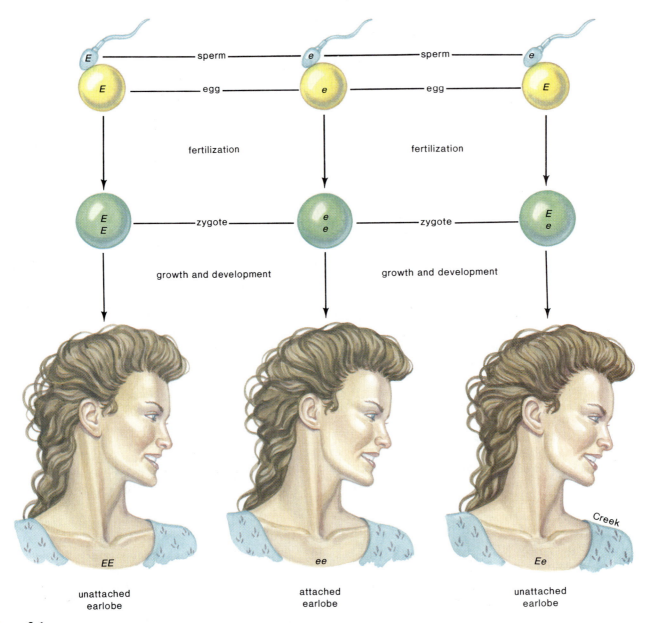

Figure 2.1

Genetic inheritance. Individuals inherit two genes for every characteristic of their anatomy and physiology. The inheritance of a single dominant gene (*E*) causes an individual to have unattached earlobes; two recessive genes (*ee*) cause an individual to have attached earlobes. Notice that each individual receives one gene from the father (by way of a sperm) and one gene from the mother (by way of an egg).

TABLE 2.1	GENOTYPE AND PHENOTYPE	
Genotype (in letters)	**Genotype (in words)**	**Phenotype**
EE	Homozygous (pure) dominant	Unattached earlobes
ee	Homozygous (pure) recessive	Attached earlobes
Ee	Heterozygous (hybrid)	Unattached earlobes

SIMPLE AUTOSOMAL INHERITANCE

Many times parents would like to know the chances of an individual child having a certain genotype and, therefore, a certain phenotype. If one of the parents is homozygous dominant (*EE*), it is obvious that the chances of a child with unattached earlobes is 100%, because this parent has only a dominant allele (*E*) to pass on to the offspring. On the other hand, if both parents are homozygous recessive (*ee*), there is a 100% chance that each child will have at-

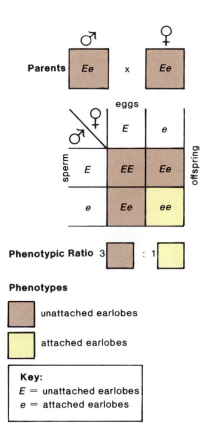

Figure 2.2

When the parents are heterozygous, each child has a 75% chance of having the dominant phenotype and a 25% chance of having the recessive phenotype.

tached earlobes. However, there are instances in which the expected phenotype is not so easily ascertained. If both parents are heterozygous, then what are the chances that a child will have unattached or attached earlobes? To solve a problem of this type, it is customary *first* to indicate the genotype of the parents and their possible gametes.

Genotypes: *Ee*   *Ee*
Gametes: *E* and *e*   *E* and *e*

Second, a Punnett Square is used to determine the phenotype ratio among the offspring when all possible sperm are given an equal chance to fertilize all possible eggs (fig. 2.2). The possible sperm are lined up along one side of the square, and the possible eggs are lined up along the other side of the square (or vice versa). The ratio among the offspring in this case is 3:1 (three children with unattached earlobes to one with attached earlobes). This means that there is a ¾ chance (75%) for each child to have unattached earlobes and a ¼ chance (25%) for each child to have attached earlobes.

Another cross of particular interest is that between a heterozygous individual (*Ee*) and a pure recessive (*ee*). In

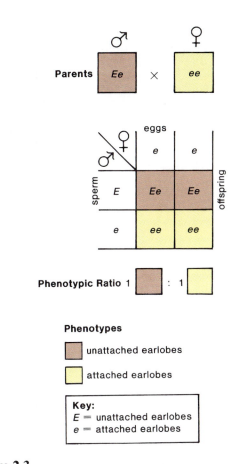

Figure 2.3

When one parent is heterozygous and the other recessive, each child has a 50% chance of having the dominant phenotype and a 50% chance of having the recessive phenotype.

TABLE 2.2	POSSIBLE MATINGS	
Genotype of Parents	**Chance of Dominant Phenotype**	**Chance of Recessive Phenotype**
$EE \times EE$	100%	0%
$EE \times Ee$	100%	0%
$EE \times ee$	100%	0%
$Ee \times Ee$	75%	25%
$Ee \times ee$	50%	50%
$ee \times ee$	0%	100%

this case, the Punnett Square operation shows that the ratio among the offspring is 1:1, and the chance of the dominant or recessive phenotype is ½, or 50% (fig. 2.3). Table 2.2 summarizes the expected results from these matings.

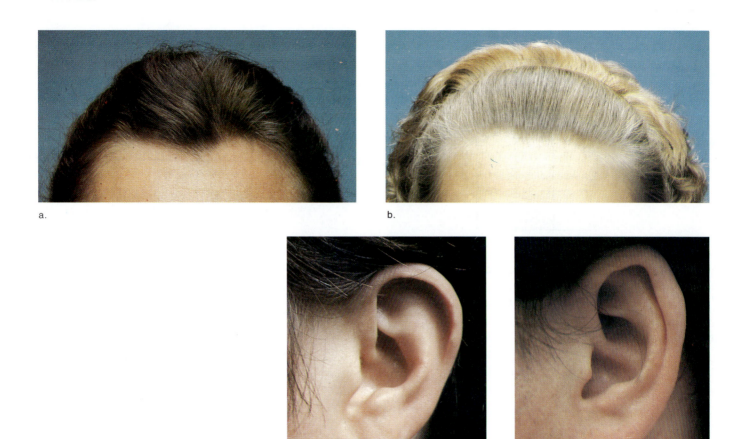

Figure 2.4
Common inherited characteristics in human beings. Widow's peak (*a*) is dominant over (*b*) continuous hairline. Unattached earlobe (*c*) is dominant over (*d*) attached earlobe.

Figure 2.4 shows some characteristics that are inherited as if they were controlled by a pair of alleles, one of which is dominant over the other. Each child has a 75% chance of having the dominant phenotype if the two parents are heterozygous and a 50% chance if one parent is heterozygous and the other is recessive. Each child has a 25% chance of having the recessive phenotype if the parents are heterozygous and a 50% chance if one parent is a heterozygous and the other is pure recessive.

SIMPLE AUTOSOMAL DISORDERS

It is now apparent that many human disorders are genetic in origin. Genetic disorders are medical conditions caused by alleles inherited from the parents. Some of these conditions are controlled by autosomal dominant or recessive alleles.

When studying human genetic disorders, biologists often construct pedigree charts to show the pattern of inheritance for a characteristic within a group of people. In a pedigree chart, males are designated by squares and females are designated by circles (fig. 2.5). Shaded circles and squares indicate affected individuals. A line between a square and a circle represents a couple who have mated. A vertical line going downward leads to a single child. (If there is more than one child, they are placed off a horizontal line.)

Autosomal Recessive Disorders

Figure 2.5 shows a typical pedigree chart for a recessive genetic disorder. It is obvious that a genetic disorder is recessive when an affected child is born to parents who appear to be normal. Other ways of recognizing a recessive disorder are also given in figure 2.5.

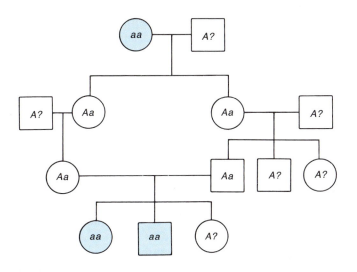

Figure 2.5

Sample pedigree chart for an autosomal recessive genetic disorder. Which individuals in this chart are known carriers? The question mark means that it is not known whether the individual is a homozygous dominant or a heterozygote.

• ─────────── HUMAN ISSUE ─────────── •

One English immigrant and her 13 children gave rise to at least 432 cases of Huntington disease in Australia. This report highlights the fact that passing on a genetic disease can have tragic consequences. Should everyone be tested for genetic disorders and then restricted from reproducing if they have the potential for passing on a serious genetic disorder? Or, should everyone be educated about genetic disorders and the possibility of being tested? Or, should we simply rely on medical remedies if a serious genetic disease does happen to be passed on? Where do you stand?

Parents with a heterozygous genotype, that is, those who carry a hidden faulty gene, are called **carriers.** When a carrier reproduces with another carrier, there is a 25% chance that any of their children will show the condition. It is important to realize that "chance has no memory," therefore, each child of these parents has a 25% chance of having the disorder. In other words, it is possible that if a heterozygous couple had four children, each child might have the condition.

Recessively inherited genetic disorders are sometimes more prevalent among members of a particular ethnic group. Members of the same ethnic group are more apt to be carriers for the same recessive disorder. Members of the same family are even more likely to be carriers for the same recessive disorder.

Cystic Fibrosis

Cystic fibrosis is the most common lethal genetic disease among Caucasians in the United States. About 1 in 20 Caucasians is a carrier, and about 1 in 2,000 children born

to this group inherits the disorder. In these children the mucus in the lungs and digestive tract is particularly thick and viscous. The presence of mucus in the lungs makes breathing extremely difficult. In the digestive tract, the thick mucus impedes the secretion of pancreatic juices, and food cannot be properly digested; large, frequent, and foul-smelling stool occurs. A few individuals have been known to survive childhood, but most die from recurrent lung infections.

In the past few years much progress has been made in understanding cystic fibrosis. First, it was discovered that water fails to enter cells, and it is believed to be the lack of water in the lungs that causes the mucus to be so thick. Second, it is now known that the gene for cystic fibrosis is located on chromosome 7. Researchers are hopeful that they will soon have the gene isolated and know how it functions.

Tay Sachs Disease

Tay Sachs is a well-known genetic disease among Jewish people in the United States, most of whom are of Central and Eastern European descent. The nerve cells of children with Tay Sachs are at first normal, but then they begin to fill up with a certain type of fatty material that the cell is unable to break down. Therefore, the Tay Sachs phenotype is not at first apparent. Development begins to slow down between 4 and 8 months of age as neurological impairment and psychomotor difficulties become apparent. The child gradually becomes blind and helpless, develops uncontrollable seizures, and eventually becomes paralyzed. There is no treatment or cure for Tay Sachs disease, and most affected individuals die by the age of three or four.

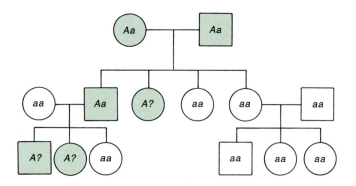

**Autosomal Dominant
Genetic Disorders**

- Affected children usually have an affected parent.
- Heterozygotes are affected.
- Two affected parents can produce an unaffected child.
- Unaffected parents do not have affected children.
- Both males and females are affected with equal frequency.

Figure 2.6

Sample pedigree chart for an autosomal dominant genetic disorder. Affected individuals are shaded. Are there any carriers? Why not?

Tay Sachs can be prevented, however, through genetic counseling. A test to detect carriers of Tay Sachs is available. Prenatal diagnosis is also possible following either amniocentesis or chorionic villi sampling (chapter 9).

Phenylketonuria (PKU)

Phenylketonuria (PKU) occurs in 1 out of 20,000 births, and so it is not as frequent as the disorders discussed thus far. When it does occur, the parents are very often closely related. Affected individuals are unable to metabolize the common amino acid phenylalanine, and the abnormal breakdown product, a phenylketone, accumulates in the urine. Newborns are routinely tested, and if they lack the necessary enzyme, they are placed on a diet low in phenylalanine. This diet must be continued until the brain is fully developed or severe mental retardation develops. If a woman who is homozygous recessive for PKU becomes pregnant she should again resume the diet, because the high level of phenylalanine in her system can pass through the placenta and may adversely affect development of the fetus.

Autosomal Dominant Disorders

Figure 2.6 shows a pedigree chart for a dominant genetic disorder. A child who has the condition must have a parent with the condition unless, of course, a **mutation** (genetic change) has occurred (chapter 4). Other ways of recognizing a dominant disorder are also given.

Usually an offspring affected with a dominant genetic disorder has one heterozygous parent and one homozygous recessive parent. Therefore, the child had a 50% chance of getting the faulty gene or escaping it completely. Again, it must be remembered that chance has no memory, and since each child has the same genetic chance it would be possible for several children in the same family to inherit a dominant genetic disease.

Neurofibromatosis (NF)

Neurofibromatosis (NF), sometimes called the elephant man[1] disease, is one of the most common genetic disorders, afflicting roughly 1 in 3,000 people, including an estimated 100,000 in the United States. It is seen equally in every racial and ethnic group throughout the world.

At birth, or perhaps later, the individual may have six or more large tan spots on the skin. Such spots may increase in size and number and get darker. Small benign tumors (lumps) called neurofibromas may occur under the skin or deeper. Neurofibromas are made up of nerve cells and other cell types.

In most cases, symptoms of NF are mild and patients live a normal life; however, in some cases the effects are severe. Skeletal deformities including a large head are seen; eye and ear tumors can lead to blindness and hearing loss. Many children with NF have learning disabilities and may be overactive.

Only recently, researchers have been able to determine that this gene is located on chromosome 17. They believe the gene to be rather large because of its varying effects and because about half of all NF cases are the result of new mutations in one of the parents.

Huntington Disease

As many as 1 in 20,000 persons in the United States have Huntington disease (HD), a neurological disorder that affects specific regions of the brain. Most individuals who inherit the allele appear normal until middle age. Then, minor disturbances in balance and coordination lead to progressively worse neurological disturbances (fig 2.7). The victim becomes insane before death occurs.

Much has been learned about Huntington disease. The gene is located on chromosome 4, and there is a test to determine if the dominant gene has been inherited. Since treatment is not available, however, few may want to have this information. Carrying on a normal life knowing that one day the symptoms of Huntington disease are expected to appear would be difficult. Also, those having the allele

---

1. Although neurofibromatosis is commonly associated with Joseph Merrick, the severely deformed nineteenth-century Londoner depicted in *The Elephant Man*, researchers today believe Merrick actually suffered from a much rarer disorder called Proteus syndrome.

Figure 2.7

After studying the chromosomes of hundreds of persons with Huntington disease, many of whom were members of related families in Venezuela, investigators determined that the gene(s) causing this genetic disease are located on chromosome 4. A test has been developed that enables investigators to analyze this chromosome in order to predict whether a person will eventually develop the adult-onset disease.

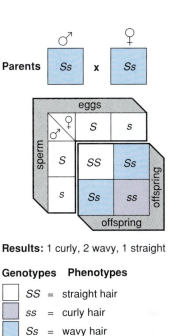

Results: 1 curly, 2 wavy, 1 straight

Genotypes		Phenotypes
☐	*SS* =	straight hair
☐	*ss* =	curly hair
☐	*Ss* =	wavy hair

Figure 2.8

Incomplete dominance. Among Caucasians, neither straight nor curly hair is dominant. When two wavy-haired individuals reproduce, each offspring has a 25% chance of having either straight or curly hair and a 50% chance of having wavy hair, the intermediate phenotype.

might see signs and symptoms in themselves when they are actually still normal.

Research is being conducted to determine the underlying cause of Huntington disease. It is known that the brain of an individual with the disorder produces more than the usual amount of an excitotoxin that can overstimulate certain neurons. This is believed to lead to the death of these neurons and the subsequent symptoms of Huntington disease. Researchers are looking for a chemical that can block the effects of the excitotoxin.

INCOMPLETE DOMINANCE

It is possible for two genes to be partially expressed, in which case both equally affect the phenotype. For example, a union between a straight-haired person and a curly haired person produces children with wavy hair:

SS = straight hair

ss = curly hair

Ss = wavy hair

(In this instance a capital and lowercase letters are used for the key. If this type of key is chosen, remember that *Ss* will have an intermediate genotype.)

Figure 2.8 shows that if two wavy-haired individuals mate, any of the three phenotypes is possible. The chances for straight hair are 25%, the chances for wavy hair are 50%, and the chances for curly hair are 25%.

Sickle-cell disease is an example of an incompletely dominant genetic disorder in which:

AA = normal

SS = sickle-cell disease

SA = sickle-cell trait

Genes and Autosomal Inheritance   **27**

(In this instance only capital letters are used in the key. If this type of key is chosen, remember that S and A are alleles.) The letter S stands for hemoglobin S and the letter A stands for hemoglobin A. Hemoglobin A is normal hemoglobin, the red respiratory pigment found in blood cells. Hemoglobin S molecules are relatively insoluble in water, and they tend to bind together to produce red cells that have a sicklelike shape (fig. 2.9).

A person with sickle-cell disease (SS) has sickle-shaped cells that cannot pass easily along small blood vessels. The sickle-shaped cells either break down or they clog blood vessels. Therefore, the individual suffers from poor circulation, anemia, and sometimes internal hemorrhaging. Jaundice, episodic pain of the abdomen and joints, poor resistance to infection, and damage to internal organs are also symptoms of sickle-cell disease.

Persons with the sickle-cell trait do not usually exhibit symptoms of the disease unless they undergo dehydration or mild oxygen deprivation. At such times, the cells become sickle-shaped, clogging their blood vessels and leading to pain and even death. A study of sudden deaths during basic training in the army showed that a person with the sickle-cell trait was 40 times more likely to die, compared to normal recruits. This has caused the Army Medical Corps to advise drill instructors to train recruits more gradually, to give them enough to drink, and to make allowances for heat and humidity when planning their workouts. Since 8% to 13% of blacks are believed to have the sickle-cell trait, the chances of a carrier reproducing with another carrier are considered to be higher than usual.

An easily performed blood test can detect those with the sickle-cell trait, and special centers have been established to test interested persons quickly and conveniently. Prenatal testing is also possible.

POLYGENIC INHERITANCE

Two or more pairs of alleles may affect the same trait, sometimes in an additive fashion. **Polygenic inheritance** can cause the distribution of human characteristics according to a bell-shaped curve, with most individuals exhibiting the average phenotype. The more genes that control the trait, the more continuous the distribution will be.

Skin Color Inheritance

Just how many pairs of alleles control skin color is not known, but a range in colors can be explained on the basis of two pairs. When a black-skinned person has children by a white-skinned person, the children have medium-

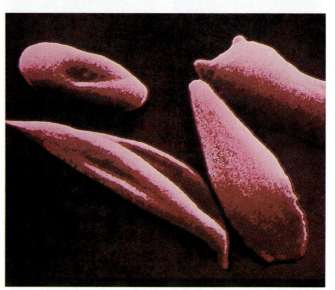

Figure 2.9
Electron micrograph of normal and sickle-shaped blood cells. Sickle-shaped blood cells are in people with sickle-cell disease.

brown skin, but two individuals with medium-brown skin can produce children who range in skin color from black to white. If we assume that two pairs of alleles control skin color, then:

Black	=	*AABB*
Dark brown	=	*AABb* or *AaBB*
Medium brown	=	*AaBb* or *AAbb* or *aaBB*
Light brown	=	*Aabb* or *aaBb*
White	=	*aabb*

Figure 2.10

Inheritance of skin color. This white-skinned husband (*aabb*) and his medium-brown-skinned wife (*AaBb*) had fraternal twins, one of whom has white and one of whom has medium-brown skin.

Figure 2.11

Jack Yale (*left*) and Oskar Stohr (*right*) are identical twins with remarkably similar behavior even though they were reared separately.

If a person with medium-brown skin reproduces with a person with white skin, the very darkest skin possible in their child is medium brown, but a white-skinned child is also possible (fig. 2.10). If one parent has only recessive alleles and the other has some dominant alleles, a child can only receive dominant alleles from that parent. Therefore, the child can be no darker than the number of dominant alleles that can be received. One must keep in mind that each parent can give only a total of two alleles and one of each kind. So, for example, it would be impossible for a union between a light-skinned person and a white-skinned person to produce a black child.

Behavioral Inheritance

Is behavior primarily inherited, or is it shaped by environmental influences? This nature (inherited) versus nurture (environment) question has been asked for a long time, and twin studies have been employed to attempt to find the answer. Twins can be identical (derived from the same fertilized egg) or fraternal (derived from two separate eggs) (chapter 9). Identical twins have inherited exactly the same chromosomes and genes, while fraternal twins have no more genes in common than do any other brother and sister.

Twin studies help determine to what extent behavior is inherited. It has been found that fraternal twins raised in the same environment are not remarkably similar in behavior, whereas identical twins raised separately are sometimes remarkably similar. For example, Oskar Stohr was raised as a Catholic by his grandmother in Nazi Ger-

many; Jack Yufe was raised by his Jewish father in the Caribbean (fig. 2.11). Yet these two men "like sweet liquers, . . . store rubber bands on their wrists, read magazines from back to front, dip buttered toast in their coffee and have highly similar personalities."[1]

Responses to a questionnaire designed to provide additional information about behavioral traits showed that identical twins reared separately tend to have more similar personalities than fraternal twins reared together. Altogether the data seemed to show that about 50% of the *differences* in human personality traits was due to polygenic inheritance and 50% was due to environmental influence.

1. C. Holden, "Identical Twins Reared Apart," *Science* 207(1980):1323–28.

Polygenic Disorders

A number of serious genetic disorders, such as a cleft lip or palate, clubfoot, congenital dislocation of the hip, and certain neural tube defects, are traditionally believed to be controlled by a combination of genes. This belief is being challenged by researchers who studied the inheritance of cleft palate in a large family in Iceland. These researchers reported the finding of a cleft palate gene on the X chromosome.

MULTIPLE ALLELES

In inheritance by **multiple alleles,** the trait is controlled, for example, by three alleles. However, each person has only two of the three possible alleles.

ABO Blood Type

Blood is a body tissue composed of two parts: cells and a liquid called plasma. When the red blood cells of one individual are mixed with the plasma of another person, the plasma frequently causes the red blood cells to clump, or agglutinate (fig. 2.12). This occurs because the membranes of red blood cells contain antigens and the plasma contains antibodies that react to antigens. The antibodies in plasma usually protect us from disease because they react to various microorganisms such as those that cause measles, polio, or strep throat. But, in fact, any foreign substance (e.g., another person's red blood cells) can be an antigen that specific antibodies will react to.

The surfaces of red blood cells have many antigens, but two of interest to us have been assigned the letters *A* and *B*. Some individuals have the antigen A and/or B on their red blood cells, and some have neither, signified by an O. Table 2.3 shows that there are four possible phenotypes but six possible genotypes for ABO blood type. Notice that the alleles *A* and *B* do not show dominance over each other—an individual who inherits both alleles will produce red blood cells with both A and B antigens and will have type AB blood. On the other hand, the *A* and *B* alleles are dominant to the recessive *O* allele; therefore, an individual who inherits the alleles *A* and *O* will produce red blood cells with only A antigens and will have type A blood.

From our study of genetics so far, it should come as no surprise that children can have a blood type different from that of either parent. For example, it is possible for

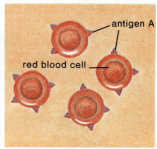
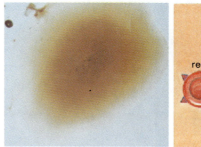

a.

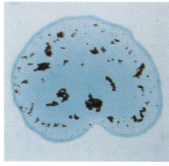
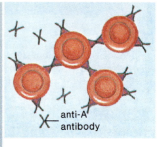

b.

Figure 2.12

a. Red blood cells in this sample carry A antigen but have not agglutinated (clumped) because anti-A antibody is absent. *b.* Red blood cells in this sample carry A antigen and have agglutinated because anti-A antibody is present.

TABLE 2.3   FREQUENCY OF BLOOD TYPES IN THE UNITED STATES

Phenotype	Genotype	Frequency among Caucasians	Frequency among Blacks
Type AB	AB	3%	3.7%
Type B	BO, BB	10%	21%
Type A	AO, AA	42%	26%
Type O	OO	45%	49.3%

parents who have blood type B to have an offspring who has blood type O if, for example, the genotype of each parent is BO (fig. 2.13*a*). However, it can also be seen that a person with blood type AB cannot have a child who has

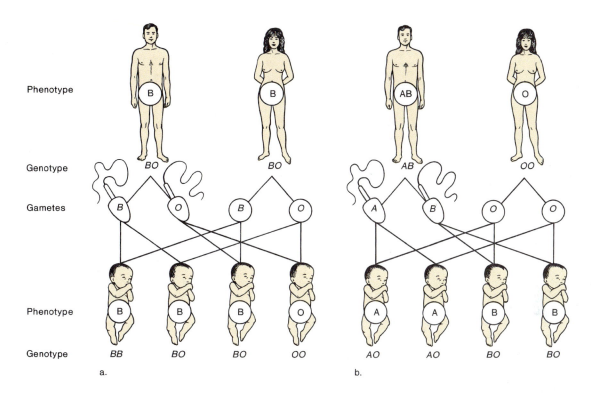

Figure 2.13

A-B-O blood type inheritance. *a.* Cross between parents, both of whom have blood type B and the genotype BO can produce a child with the blood type B or blood type O. Note the genotype of each possible offspring. *b.* Cross between a male who has blood type AB and the genotype AB and a female who has blood type O and the genotype OO can produce children with the blood type A or B but not O. Note the genotypes of the offspring.

type O blood even if the other parent has type O blood (fig. 2.13*b*). Obviously a person who has blood type O cannot have a child who has blood type AB. Such considerations as these can be used in cases of disputed paternity to decide if a person is or is not the parent of a child.

Rh Factor

Most Americans have blood type O^+. The positive sign means that their red blood cells contain an antigen called the Rh factor. The Rh factor is inherited as a simple dominant allele in which D = Rh factor and d = absence of the factor. Therefore, it is possible for parents with Rh positive (Rh^+) blood to have a child with Rh negative (Rh^-) blood if the parents are heterozygous.

The Rh factor is of particular concern when the woman has Rh^- blood and the man has Rh^+ blood. In such

a case, a child may, of course, have Rh^+ blood. As illustrated in figure 2.14, a pregnant woman with Rh^- blood may react to the blood cells of an Rh^+ child by producing anti-Rh antibodies. This will occur just before or at the time of delivery, when some of the child's blood cells can enter her system. If the woman becomes pregnant with another Rh^+ child, these antibodies may cross the placenta and cause destruction of the unborn child's red blood cells, causing a condition called hemolytic disease of the newborn (HDN). Years ago, the only available treatment for HDN was an immediate blood transfusion for the newborn. Today, however, prevention is possible because an Rh^- woman can receive an injection (called Rho Gam) after the birth of an Rh^+ child. The injection will prevent her from producing antibodies against Rh^+ blood cells. ■

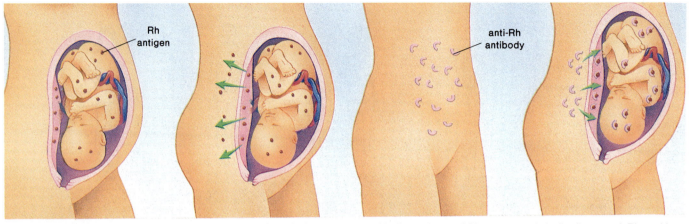

a. Child is Rh positive
   Mother is Rh negative

b. Red blood cells leak
   across placenta

c. Mother makes anti-Rh
   antibodies

d. Antibodies attack Rh-positive
   red blood cells

Figure 2.14

Hemolytic disease of the newborn (HDN). *a.* Baby is Rh positive (Rh$^+$) because the red blood cells carry the Rh antigen; mother is Rh negative (Rh$^-$) because her red blood cells do not carry the antigen. *b.* Some of the baby's cells escape into the mother's system. *c.* Mother begins manufacturing anti-Rh antibodies. *d.* During a subsequent pregnancy, mother's anti-Rh antibodies cross placenta to destroy the new baby's red blood cells if baby is Rh positive (Rh$^+$).

SUMMARY

Genes located within the chromosomes control the characteristics of an individual. A gene normally has two alleles. A dominant allele is assigned a capital letter and a recessive allele is assigned a lowercase letter. An individual with the genotype *Ee* would have the phenotype of unattached earlobes.

A Punnett Square can help to determine the phenotype ratio among offspring because all possible sperm types of a particular father are given an equal chance to fertilize all possible egg types of a particular mother. When a heterozygous individual reproduces with another heterozygote there is a 75% chance the offspring will have the dominant phenotype and a 25% chance the child will have the recessive phenotype. When a heterozygote reproduces with a pure recessive, the offspring has a 50% chance of either phenotype.

Determination of inheritance is sometimes complicated by the fact that characteristics are controlled by partially dominant alleles, more than one pair of alleles (polygenic inheritance), or by multiple alleles. Skin color illustrates polygenic inheritance and ABO blood type illustrates inheritance by multiple alleles.

It is now known that many disorders of humans are genetic in origin.

KEY TERMS

alleles 21

carriers 25

cystic fibrosis 25

dominant allele 21

genotype 21

heterozygous 21

homozygous dominant 21

homozygous recessive 21

multiple alleles 30

mutation 26

neurofibromatosis (NF) 26

phenotype 21

phenylketonuria (PKU) 26

polygenic inheritance 28

recessive allele 21

sickle-cell disease 27

Tay Sachs 25

REVIEW QUESTIONS*

Assume the trait is dominant for the following questions:

1. A woman heterozygous for polydactyly, a condition that produces six fingers and toes, is married to a man without this condition. What are the chances that her children will have six fingers and toes?

2. A young man's father has just been diagnosed as having Huntington disease. What are the probable chances that the son will inherit this condition?

3. Black hair is dominant over blond hair. A woman with black hair whose father had blond hair reproduces with a blond-haired man. What are the chances of this couple having a blond-haired child?

4. Your maternal grandmother Smith had Huntington disease. Aunt Jane, your mother's sister, also had the disease. Your mother dies at age 75 with no signs of Huntington disease. What are your chances of getting the disease?

5. Could a person who can curl her tongue have parents who cannot curl their tongues? Explain your answer.
   *Answers appear on page 34.

Assume the trait is recessive for the following questions:

6. Parents who do not have Tay Sachs produce a child who has Tay Sachs. What is the genotype of each parent? What are the chances each child will have Tay Sachs?

7. One parent has lactose intolerance, the inability to digest lactose, the sugar found in milk, and the other is heterozygous. What are the chances that their child will have lactose intolerance?

8. A child has cystic fibrosis. His parents are normal. What is the genotype of all persons mentioned?

Assume the trait is incompletely dominant for the following questions:

9. What are the chances that a person pure for straight hair who is married to a person pure for curly hair will have children with wavy hair?

10. One parent has sickle-cell disease and the other is perfectly normal. What are the phenotypes of their children?

11. A child has sickle-cell disease but her parents do not. What is the genotype of each parent?

12. Both parents have the sickle-cell trait. What are their chances of having a perfectly normal child?

Assume the trait is controlled by multiple alleles, or more than one pair of alleles for the following problems:

13. The genotype of a woman with type B blood is *BO*. The genotype of her husband is *AO*. What could be the genotypes and phenotypes of the children?

14. A man has type AB blood. What is his genotype? Could this man be the father of a child with type B blood? If not, why not? If so, what blood types could the child's mother have?

15. Baby Susan has type B blood. Her mother has type O blood. What type blood could her father have?

16. Fill in the following pedigree chart to give the probable genotypes of the twins pictured in figure 2.10.

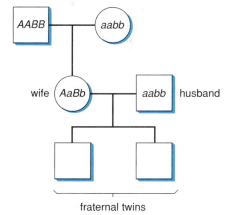

fraternal twins

CRITICAL THINKING QUESTIONS

1. In fruit flies, long wings are dominant to short wings. A student is observing the results of a cross and counts 300 long-winged flies and 98 short-winged flies. (a) What was the genotype of the parent flies? How do you know? (b) Of the long-winged flies how many do you predict are homozygous dominant and how many are heterozygous? Why do you say so?

2. Persons with sickle-cell disease tend to die at a young age and leave no offspring. Why doesn't sickle-cell disease cease to be a threat in future generations?

3. A woman with white skin has parents with medium-brown skin. If this woman reproduces with a black man, what is the darkest skin color possible for their children? The lightest skin color possible? What is your reasoning?

4. If a woman with genotype *BODd* reproduces with a person having genotype *AOdd,* what blood types might their children have? What is your reasoning? Could this mating be one in which hemolytic disease of the newborn might occur? Why or why not?

ANSWERS TO REVIEW QUESTIONS

1. 50%
2. 50% (most likely father is heterozygous)
3. 50%
4. None
5. No. Only homozygous recessives cannot curl the tongue, and persons with this genotype cannot pass on the ability to curl the tongue.
6. *Tt,* 25%
7. 50%
8. Child *aa,* parents *Aa*
9. 100%
10. sickle-cell trait
11. *Ss*
12. 25%
13. *BO* (type B blood), *AO* (type A blood), *AB* (type AB blood), *OO* (type O blood)
14. *AB.* He could be the father of a child with type B if mother is *BB, BO,* or *OO.*
15. Father could be *BB* or *BO* or *AB.*

Sex-Linked Inheritance

The sex chromosomes contain genes just as the autosomal chromosomes do. Some of these genes determine whether the individual is a male or female. Investigators have found a gene they call the testis-determining factor (TDF) in the *sex-determining region* of the Y chromosome. When this gene is lacking from the Y chromosome, the individual is a female even though the chromosomal inheritance is XX. Although this gene is normally found on the Y chromosome it sometimes occurs on the X chromosome because of a chromosome mutation. If the gene is present on an X chromosome an XX individual is a male.

Another gene that has been found on the Y chromosome is called H-Y$^+$. This allele directs the synthesis of the H-Y antigen that is present in the cell membrane of almost all cells in males but not in those of a female. It is called an antigen because females produce antibodies against it. To test for maleness, it is possible to suspend a sample of white blood cells in a solution that contains some of these antibodies. If the cells carry the H-Y antigen, indicating the person is a male, the antibodies bind with them.

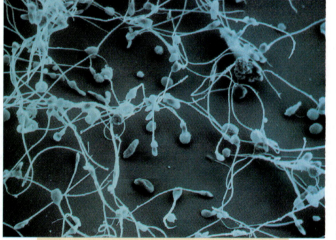

Human sperm following ejaculation.

SEX-LINKED GENES

Genes on the sex chromosomes are said to be **sex linked;** a gene that is only on the X chromosome is **X linked,** and a gene that is only on the Y chromosome is **Y linked.** Most sex-linked genes are on the X chromosome and the Y chromosome is blank for these genes. Very few genes have been found on the Y chromosome, as you might predict, since it is much smaller than the X chromosome.

The X chromosomes carry many genes unrelated to the sex of the individual, and we will look at a few of these in depth. It would be logical to suppose that a sex-linked trait is passed from father to son or from mother to daughter, but this is not the case. A male always receives a sex-linked condition from his mother, from whom he inherited an X chromosome. The Y chromosome from the father does not carry an allele for the trait. Usually the trait is recessive; therefore, a female must receive two alleles, one from each parent before she has the condition.

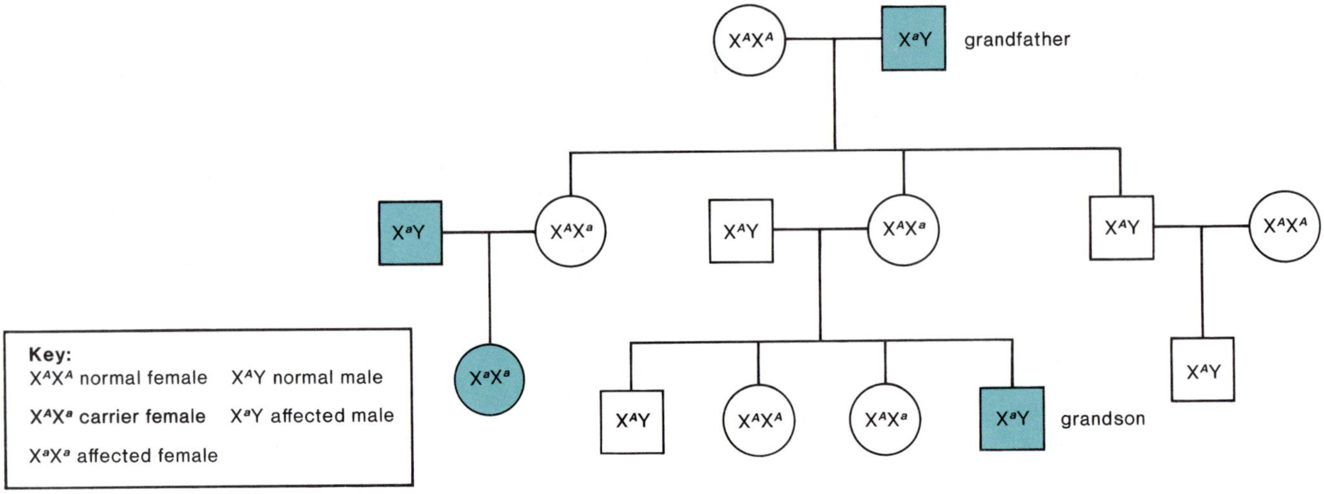

Figure 3.1
Sample pedigree chart for an X-linked recessive genetic disorder in which affected individuals are shaded more heavily than carriers.

X-Linked Genes

When examining X-linked traits, the allele on the X chromosome appears as a letter attached to the X chromosome. Since most sex-linked conditions are recessive, the expected key is as follows:

X^A = normal
X^a = abnormal trait

The possible genotypes in both males and females are as follows:

$X^A X^A$ = a normal female
$X^A X^a$ = a carrier female
$X^a X^a$ = a female with the trait
$X^A Y$ = a normal male
$X^a Y$ = a male with the trait

Note that the second genotype is a carrier female because although a female with this genotype appears normal, she is capable of passing on an allele for an X-linked characteristic.

From the key you would deduce that when a gene is carried on an X chromosome, females would have two active alleles for the trait in each of their cells. This is not the case, as explained in the reading, "X Chromosomes in Female Body Cells."

Pedigree Chart

Figure 3.1 shows a pedigree chart for an X-linked condition. Affected individuals are darkly colored and carrier females are lightly colored. Notice the following:

More males than females are affected because if a male inherits the recessive allele, there can be no corresponding dominant allele to offset it.

Affected males can have parents who are not affected.

Affected females must have a father who is affected and a mother who is affected or is a carrier because affected females have to receive a recessive allele from both parents.

The trait often skips a generation from grandfather to grandson because women are carriers of the condition.

If the woman is affected, all her sons will be affected.

SEX-LINKED DISORDERS

Several X-linked disorders are known, and we will study several of these.

Males inherit only one X chromosome, while females have two X chromosomes. Most sex-linked genes are on the X chromosome. Does this mean that females have two alleles for each sex-linked gene in their cells? Or, is there some sort of dosage compensation so that males and females are equal in terms of active alleles?

Years ago, M. L. Barr observed a consistent difference between nondividing cells taken from females and males. Females, but not males, have a small, dark-staining spot in their nuclei, which is called a Barr body after its discoverer (fig. 3.A). In 1961, Mary Lyon, a British geneticist, proposed that the Barr body is a condensed, inactive X chromosome. As it turns out, one of the X chromosomes is inactivated in the cells of female embryos, but which one of the two is determined by chance? About 50% of the cells have one X chromosome active, and 50% have the other X chromosome active. The female body, therefore, is a mosaic, with "patches" of genetically different cells. For example, human females who are heterozygous for an X-linked recessive form of ocular albinism have patches of pigmented and nonpigmented cells at the back of the eye. Women heterozygous for Duchenne muscular dystrophy have patches of normal muscle tissue and degenerative muscle tissue (the normal tissue increases in size and strength to make up for the defective tissue); women who are heterozygous for hereditary absence of

Barr body

a.

Figure 3.A

b.

a. Electron micrograph illustrates that a Barr body is a dark-staining spot that can be viewed microscopically. *b.* The coat of a calico cat is orange and black. The alleles for coat color occur at a gene locus on the X chromosome. Presumably, the black patches occur where cells have Barr bodies carrying the allele for orange, and the orange patches occur where cells have Barr bodies carrying the allele for black coat. The white areas are due to a separate gene.

sweat glands have patches of skin lacking sweat glands.

The female calico cat (see fig. 3.B) also provides phenotypic proof of the Lyon hypothesis. In these cats, an allele for black coat color is on one X chromosome, and a corresponding allele for orange coat color is carried on the other X chromosome. The patches of black and

orange in the coat can be related to which X chromosome is in the Barr bodies of the cells found in the patches.

The existence of Barr bodies also explains why metafemales and persons with Klinefelter syndrome do not show a greater degree of abnormality than they do. The extra X chromosomes form Barr bodies.

Red-Green Color Blindness

About 6% of men in the United States are **color blind** due to an inability to see green colors, and about 2% are color blind due to inability to see red colors. Both of these genes are sex linked.

If a carrier woman is married to a man with normal vision, what are their chances of having a color-blind daughter? a color-blind son?

Parents: $X^B X^b \times X^B Y$

Inspection indicates that all daughters will have normal color vision because they will all receive an X^B from their father. The sons, however, have a 50% chance of being color blind, depending on whether they receive an X^B or X^b from their mother. The inheritance of a Y from their father cannot offset the inheritance of an X^b from their mother. Figure 3.2 illustrates the use of the Punnett Square in solving sex-linked problems.

Other Recessive X-Linked Genetic Disorders

Hemophilia

There are about 100,000 hemophiliacs in the United States. The most common type of **hemophilia** is hemophilia A, due to the absence, or minimal presence, of a particular clotting factor called factor VIII. Hemophilia is called the bleeder's disease because the afflicted person's blood is unable to clot. Although hemophiliacs do bleed externally after an injury, they also suffer from internal bleeding, particularly around joints. Hemorrhages can be checked with transfusions of fresh blood (or plasma) or concentrates of the clotting protein. Unfortunately, some hemophiliacs have contracted AIDS after using whole blood and the concentrate.

At the turn of the century, hemophilia was prevalent among the royal families of Europe. All of the afflicted males could trace their ancestry to Queen Victoria of England (fig. 3.3). Since none of Queen Victoria's forebearers or relatives was afflicted, it seems that the gene she carried arose by mutation either in Victoria or in one of her parents. Her carrier daughters, Alice and Beatrice, introduced the gene into the ruling houses of Russia and Spain. Alexis, the last heir to the Russian throne before the Russian Revolution, was a hemophiliac. The present British royal family has no hemophiliacs because Victoria's eldest son, King Edward VII, did not receive the gene and therefore could not pass it on to any of his descendents.

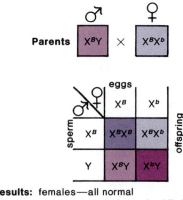

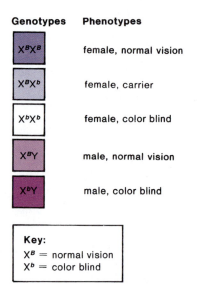

Results: females—all normal
males—1 normal: 1 color blind

Genotypes	Phenotypes
$X^B X^B$	female, normal vision
$X^B X^b$	female, carrier
$X^b X^b$	female, color blind
$X^B Y$	male, normal vision
$X^b Y$	male, color blind

Key:
X^B = normal vision
X^b = color blind

Figure 3.2

Cross involving an X-linked gene. The male parent is normal, but the female parent is a carrier; an allele for color blindness is located on one of her X chromosomes. Therefore, each son stands a 50% chance of being color blind.

HUMAN ISSUE

Duchenne muscular dystrophy is the most common lethal X-linked disorder. It occurs in 1 out of every 7,000 male births. What would you do if you were a male in love with a woman with this disorder in her pedigree chart? Not marry the woman? Not have children if the woman tests positive for being a carrier? Use the egg selection technique described in the chapter 9 reading, "Detecting Birth Defects" and have only female children? Leave it up to chance whether a boy with DMD is born?

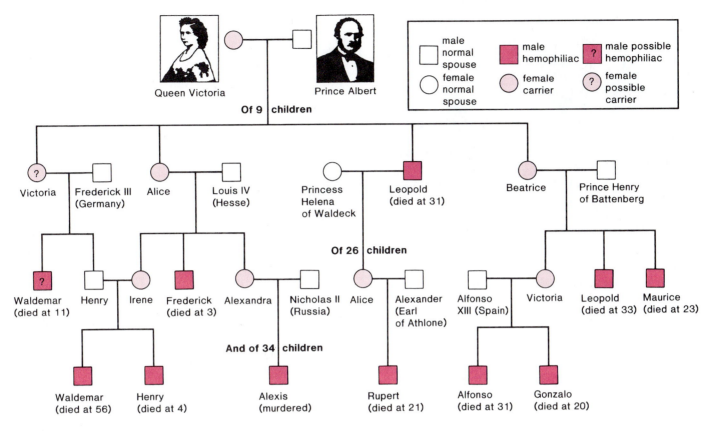

Figure 3.3

A simplified pedigree showing the X-linked inheritance of hemophilia in European royal families. Because Queen Victoria was a carrier, each of her sons had a 50% chance of having the disease and each of her daughters had a 50% chance of being a carrier. This pedigree shows only the affected individuals. Many others are unaffected, such as the members of the present British royal family.

Muscular Dystrophy

Muscular dystrophy, as the name implies, is characterized by wasting away of the muscles. The most common form, Duchenne muscular dystrophy (DMD), is X linked and occurs in about 1 out of every 7,000 male births. Symptoms such as waddling gait, toe walking, frequent falls, and difficulty in rising may appear as soon as the child starts to walk. Muscle weakness intensifies until the individual is confined to a wheelchair. Death usually occurs during the teenage years.

When the gene for muscular dystrophy was isolated, it was discovered that the absence of a protein, now called dystrophin, is the cause of the disorder. Without this protein the muscles malfunction and begin to break down. When the body attempts to repair the tissue, the formation of fibrous tissue occurs, and this cuts off the blood supply so that more and more cells die. A test is now available to detect carriers for Duchenne muscular dystrophy.

Lesch-Nyhan Syndrome

Males who inherit **Lesch-Nyhan syndrome** are mentally retarded and practice self-mutilation by chewing their lips and fingertips. The individual shows scars resulting from this practice. Curiously, the metabolic abnormality consists of accumulatin of uric acid in the blood.

Severe Combined Immune Deficiency Syndrome

Persons who have **severe combined immune deficiency syndrome (SCID)** are unable to manufacture antibodies and, therefore, are susceptible to repeated infections. To prevent this occurrence, they sometimes must spend their lives in a protective sterile bubble or a protective suit (fig. 3.4); therefore, the condition is sometimes called the bubble baby disease. Normally, the lymphocytes, a type of white blood cell, produce antibodies that fight infection. Since a person with SCID is unable to make antibodies, this natural protection against infection is an impossibility.

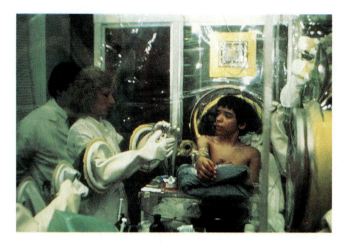

Figure 3.4

This boy had severe combined immune deficiency syndrome (SCID), an X-linked genetic disorder. He lived in a germ-free bubble for 12 years because his body failed to produce antibodies to fight infections. It was hoped that a bone marrow transplant would give him the capability to produce antibodies so that he could leave his bubble home. After receiving a marrow transplant from his sister, he was removed from his bubble for treatment, but he lived for a brief 2 weeks. His sister's bone marrow contained a cancer-causing virus he was unable to combat.

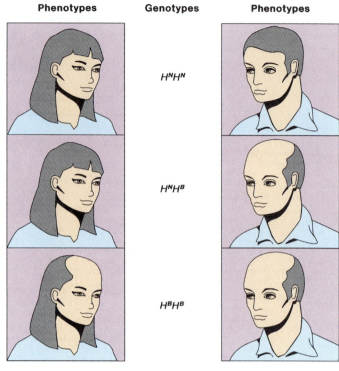

Phenotypes	Genotypes	Phenotypes
	$H^N H^N$	
	$H^N H^B$	
	$H^B H^B$	

H^N—normal hair growth
H^B—pattern baldness

Figure 3.5

Baldness is a sex-influenced characteristic. The presence of only one allele for baldness causes the condition in the male, whereas the condition does not occur in the female unless she possess both alleles for baldness.

SEX-INFLUENCED TRAITS

Not all traits we associate with the sex of the individual are due to sex-linked genes. Some are simply **sex-influenced traits.** Sex-influenced traits are characteristics that often appear in one sex but only rarely appear in the other. It is believed that these traits are governed by genes turned on or off by hormones. For example, the secondary sex characteristics, such as the beard of a male and the breasts of a female, probably are controlled by hormone balance.

Baldness is believed to be caused by the male sex hormone testosterone because males who take the hormone to increase masculinity begin to lose their hair (fig. 3.5). A more detailed explanation has been suggested by some investigators. It has been reasoned that due to the effect of hormones, males require only one gene for the trait to appear, whereas females require two genes. In other words, the gene acts as a dominant in males but as a recessive in females. This means that males born to a bald father and a mother with hair at best would have a 50% chance of going bald. Females born to a bald father and a mother with hair at worst would have a 25% chance of going bald.

Another sex-influenced trait of interest is the length of the index finger. In women the index finger is at least equal to if not longer than the fourth finger. In males the index finger is shorter than the fourth finger. ■

SUMMARY

Sex-linked genes are located on the sex chromosomes, and most alleles are X linked because they are on the X chromosome. Since the Y chromosome does not contain these alleles, only the allele received from the mother determines the phenotype in males. This means that males are more apt to display a trait controlled by an X-linked recessive allele. It also indicates that if a female does have the phenotype, then her father must also have it. Usually X-linked traits skip a generation and go from maternal grandfather to grandson by way of a carrier daughter.

Genetic disorders are sometimes X linked. Those of particular interest are hemophilia, muscular dystrophy, Lesch-Nyhan syndrome, and severe combined immune deficiency syndrome.

Some conditions are sex influenced and are apt to appear more often in one sex than the other. Sex-influenced traits are controlled by genes that are believed to be turned on and off by hormones.

KEY TERMS

color blind 38

hemophilia 38

Lesch-Nyhan syndrome 39

muscular dystrophy 39

severe combined immune deficiency
  syndrome (SCID) 39

sex-influenced traits 40

sex linked 35

X linked 35

Y linked 35

REVIEW QUESTIONS*

1. A boy has severe combined immune deficiency syndrome. What are the genotypes of the parents, who have the normal phenotype?

2. A woman is color blind and her spouse has normal vision. If they produce a son and a daughter, which child will be color blind?

*3. If a female who carries an X-linked allele for Lesch-Nyhan syndrome reproduces with a normal man, what are the chances that male children will have the condition? that female children will have the condition?

4. A girl has hemophilia. What is the genotype of her father? What is the genotype of her mother, who has a normal phenotype?

5. Are the traits (shaded circles and squares) in the following pedigree charts autosomal dominant, autosomal recessive, or sex-linked recessive traits?

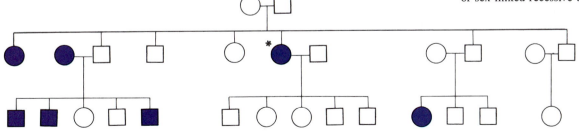

a.

*What is the genotype of this person?

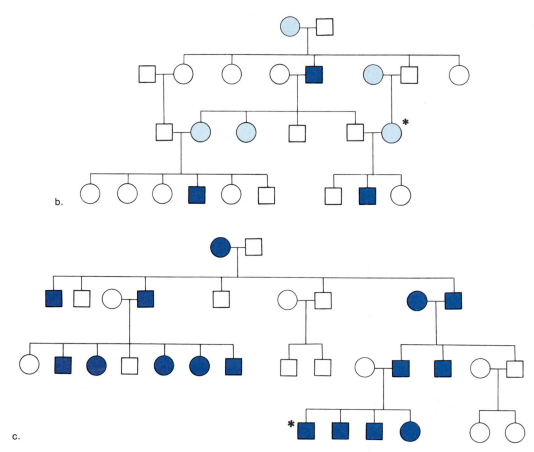

b.

c.

*What is the genotype of this person?

CRITICAL THINKING QUESTIONS

1. A male is color blind and has a continuous hairline. Are the genes for these traits on the same chromosome or different chromosomes? How do you know?

2. What is the genotype of the male in question 1?

3. If the man in questions 1 and 2 reproduces with a woman who is pure dominant for normal vision and widow's peak, what will be the genotype and phenotype of the children?

*ANSWERS TO REVIEW QUESTIONS

1. father: X^sY; mother: X^sX^s
2. son
3. 50%, none
4. X_HX_h, X_HY
5. a. autosomal recessive, aa
   b. Sex-linked recessive, X^AX^a
   c. autosomal dominant, Aa

Biochemical Genetics

Our approach to genetics thus far has been to consider the genes as particulate units of a chromosome. In contrast, biochemical genetics considers the chemical nature of the gene and the biochemical function of genes in the cell.

DNA AND RNA

Genes are made up of a chemical called **DNA (deoxyribonucleic acid).** This means that DNA is the genetic material, and that chromosomes contain DNA. In fact, DNA is found principally in the nucleus of a cell (fig. 4.1). In contrast, **RNA (ribonucleic acid)** is found both in the nucleus and the cytoplasm. A type of RNA called ribosomal RNA (rRNA) is found within the **ribosomes,** which are small structures whose subcomponents are assembled at the nucleolus within the nucleus.

Ovulation is occurring.

Later, ribosomes are found on rough endoplasmic reticulum within the cytoplasm (see also fig. 1.1).

Structure of DNA

DNA is a type of nucleic acid, and like all nucleic acids, it is formed by the sequential joining of smaller molecules called nucleotides. Nucleotides, in turn, are composed of three molecules—a *phosphate,* a *sugar,* and a *base.* The sugar in DNA is deoxyribose, which accounts for the chemical's name, deoxyribonucleic acid. There are four different nucleotides in DNA because there are four bases: **adenine (A), thymine (T), cytosine (C),** and **guanine (G)** (fig. 4.2).

When nucleotides join together, the sugar and the phosphate molecules become a backbone and the bases project to the side. In DNA, there are two such chains of nucleotides held together by weak bonds between the bases; consequently, DNA is double-stranded. Each base is bonded to another particular base, called its **complementary base**—adenine (A) is always paired with thymine (T), and cytosine (C) is always paired with guanine (G). The reverse of these statements is, of course, also true. The dotted lines in figure 4.3*a* represent the weak bonds between the bases. The structure of DNA is said to resemble a ladder, because the sugar-phosphate molecules appear to make up the sides of a ladder and the bases seem to make the steps. The ladder structure of DNA twists to form a spiral staircase called a **double helix** (fig 4.3*b*). DNA is most often shown in this double helix form.

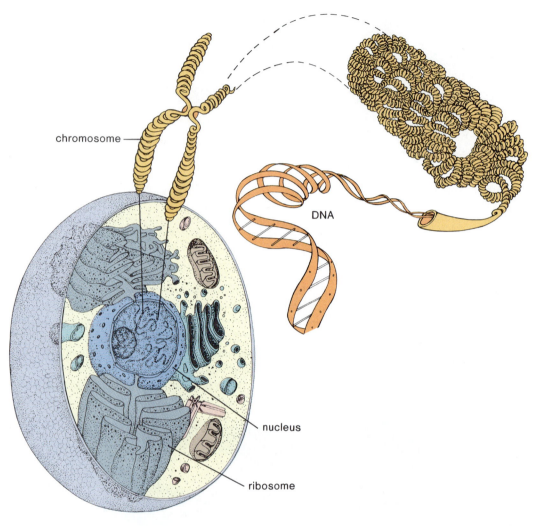

Figure 4.1

Chromosome structure. Chromosomes that are highly compacted and coiled when visible actually contain a molecule called DNA. DNA has a double helix structure that resembles a twisted ladder—note the sides and rungs of the ladder structure. In this chapter, we will also be interested in other parts of the cell, such as the ribosomes that are composed of the chemical RNA. Both DNA and RNA are molecules called nucleic acids.

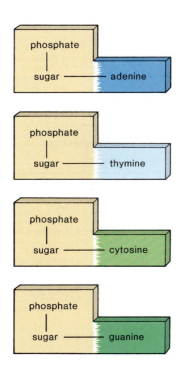

Figure 4.2

DNA structure. DNA contains four different kinds of nucleotides, molecules that in turn contain a phosphate, a sugar, and a base. The base of a DNA nucleotide can be either adenine (A), thymine (T), cytosine (C), or guanine (G).

Structure of RNA Compared to DNA

RNA is a nucleic acid made up of nucleotides containing the sugar ribose. This sugar accounts for its scientific name, ribonucleic acid. The nucleotides that compose RNA also have four possible bases: adenine (A), uracil (U), cytosine (C), and guanine (G) (fig. 4.4). In RNA, the base uracil replaces the base thymine.

RNA, unlike DNA, is always single-stranded. Similarities and differences between these two nucleic acid molecules are given in table 4.1.

Functions of DNA

The two primary functions of DNA are replication and control of protein synthesis.

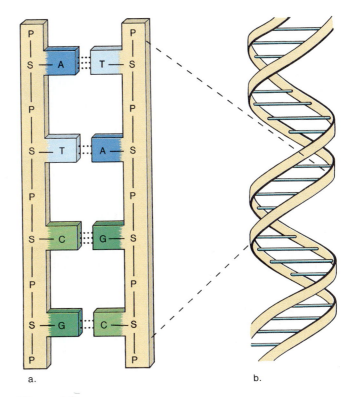

Figure 4.3

DNA is double stranded. *a.* When the DNA nucleotides join, they form two strands so that the structure of DNA resembles a ladder. The phosphate (P) and sugar (S) molecules make up the sides of the ladder, and the bases make up the rungs of the ladder. Each base is weakly bonded (dotted lines) to its complementary base (T is bonded to A and vice versa; G is bonded to C and vice versa). *b.* The DNA ladder twists to give a double helix. Each chromatid of a duplicated chromosome contains one double helix.

Replication

Between cell divisions, the daughter chromosomes must duplicate before cell division can occur again. Actually, when chromosomes duplicate, DNA is replicating, that is, it is making a copy of itself. **Replication** has been found to require the following steps:

1. The two strands that make up DNA become "unzipped" (i.e., the weak bonds between the paired bases break).
2. New complementary nucleotides, always present in the nucleus,[1] move into place beside each old strand by the process of base pairing.

---

1. The food we eat is digested into molecules such as nucleotides and amino acids, which are carried in the bloodstream to the cells.

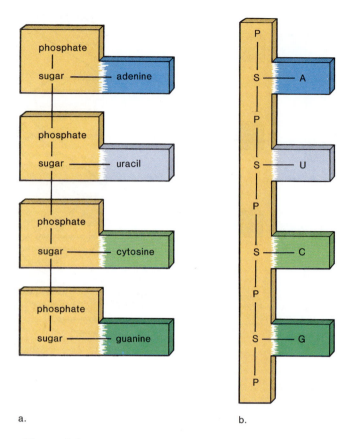

a.   b.

Figure 4.4

RNA structure. *a.* The four nucleotides in RNA each have a phosphate (P) molecule; the sugar (S); ribose; and a base that may be either adenine (A), uracil (U), cytosine (C), or guanine (G). *b.* RNA is single-stranded. The sugar and phosphate molecules join to form a single backbone, and the bases project to the side.

3. These new complementary nucleotides become joined together.
4. When the process is finished, two complete DNA molecules are present, identical to each other and to the original molecule.

Each new double helix is composed of an old and a new strand (fig. 4.5). The new strand is complementary to the old strand. Because of this, it is said that each strand of DNA serves as a *template,* or mold, for the production of a complementary strand. The entire replication process is called *semiconservative* because old strands are always paired with new strands. Because of semiconservative replication, the new double helix molecules are identical to each other and to the original molecule.

TABLE 4.1   DNA-RNA SIMILARITIES AND DIFFERENCES

DNA-RNA Similarities

Both are nucleic acids
Both are composed of nucleotides
Both have a sugar-phosphate backbone
Both have four different type bases

DNA-RNA Differences

DNA	RNA
Found in nucleus	Found in nucleus and cytoplasm
The genetic material	Helper to DNA
Sugar is deoxyribose	Sugar is ribose
Bases are A, T, C, G	Bases are A, U, C, G
Double-stranded	Single-stranded
Transcription (to give mRNA)	Translation (to give proteins)

Protein Synthesis

The protein synthesis function of DNA will be discussed shortly in some detail because it is a rather complicated process. Because DNA controls protein synthesis it controls the structure and function of the cell, as we shall see. In order to bring about protein synthesis, DNA requires the help of RNA.

Functions of RNA

RNA is a helper to DNA so that protein synthesis occurs in the manner DNA directs. DNA always resides in the nucleus, but RNA is found both in the nucleus and in the cytoplasm. This means that RNA is more intimately involved in protein synthesis because protein synthesis takes place at the ribosomes in the cytoplasm.

There are three types of RNA, each with a specific function in protein synthesis:

Ribosomal RNA (rRNA)—found within the ribosomes where protein synthesis occurs. Ribosomes are very often attached to a membrane called the endoplasmic reticulum, which ramifies throughout the cytoplasm.

Messenger RNA (mRNA)—carries information to the cytoplasm, which directs the synthesis of proteins.

Transfer RNA (tRNA)—also sees to it that proteins are made in the manner directed by mRNA.

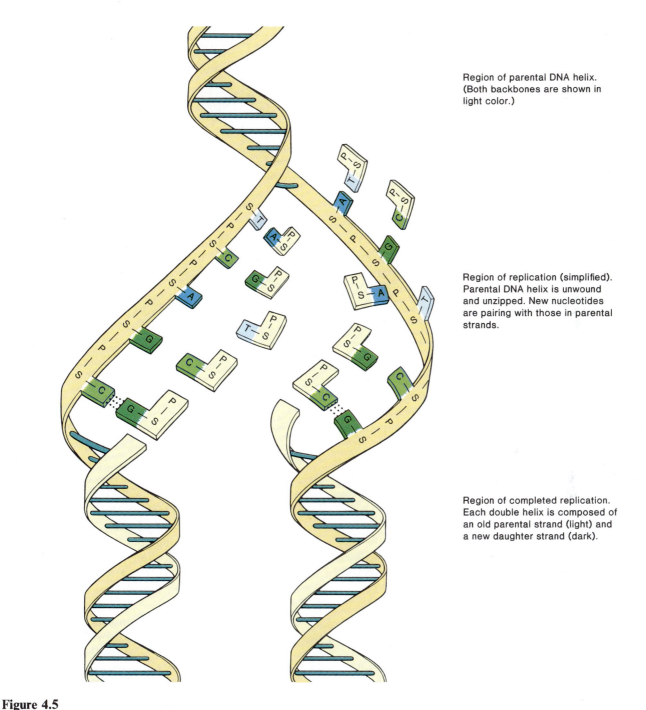

Region of parental DNA helix. (Both backbones are shown in light color.)

Region of replication (simplified). Parental DNA helix is unwound and unzipped. New nucleotides are pairing with those in parental strands.

Region of completed replication. Each double helix is composed of an old parental strand (light) and a new daughter strand (dark).

Figure 4.5

DNA replication requires these steps: (1) top—the two strands of DNA become unzipped (the weak bonds between the paired bases break); (2) upper middle—new complementary nucleotides move into place and bond with nucleotides in the old strand, (3) lower middle—new complementary nucleotides become joined together; (4) bottom—when process is finished two complete DNA double helix molecules are present. Note that DNA replication is semiconservative because each new double helix is composed of an old parental strand and a new daughter strand.

PROTEINS AND PROTEIN SYNTHESIS

Before describing the steps needed for protein synthesis, it is necessary to understand the structure and function of proteins.

Structure and Function of Proteins

Proteins are found in all cells. The protein hemoglobin is responsible for the red color of red blood cells. The antibodies are proteins, and there are other proteins in the blood as well. Muscle cells have much protein, giving them substance and their ability to contract.

Certain proteins are **enzymes,** which speed up chemical reactions in cells. Actually, cells are chemical factories in which reactions are constantly and quickly occurring even though the body has a relatively low temperature. Reactions occur quickly in cells because of the presence of enzymes. The reactions in cells form chemical or metabolic pathways. One such pathway could be represented as follows:

$$A \xrightarrow{E_A} B \xrightarrow{E_B} C \xrightarrow{E_C} D \xrightarrow{E_D} E$$

In this pathway the letters are molecules and the notations over the arrows are enzymes: molecule A becomes molecule B, and enzyme E_A speeds up the reaction; molecule B becomes molecule C, and enzyme E_B speeds up the reaction, and so forth. Notice that each reaction in the pathway has its own enzyme: enzyme E_A can only convert A to B; enzyme E_B can only convert B to C; and so forth. For this reason, enzymes are said to be *specific.*

Proteins are very large chemical molecules composed of individual units called **amino acids.** An amino acid joined to an amino acid joined to an amino acid, until there are perhaps hundreds of amino acids, results in a protein. Twenty different amino acids are commonly found in proteins. Proteins differ because the number and order of the amino acids differ (fig. 4.6). When DNA directs protein synthesis, it directs the order of the amino acids in a particular protein. DNA can do this because every three bases in DNA code for, or represent, one amino acid (table 4.2). Therefore, it is said that DNA contains a triplet code.

Protein Synthesis

Protein synthesis requires two steps, called transcription and translation. It is helpful to recall that **transcription** is often used to signify making a copy of certain information, while **translation** means to put this information into another language.

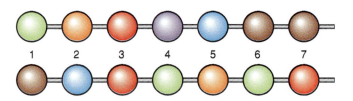

Figure 4.6

Amino acid sequences from two different proteins are represented. Proteins differ in the number and sequence of their amino acids.

TABLE 4.2	DNA CODE AND RNA CODON		
DNA Code	**mRNA Codon**	**tRNA Anticodon**	**Amino Acid (designated by color)**
TTT	AAA	UUU	
TGG	ACC	UGG	
CCG	GGC	CCG	
CAT	GUA	CAU	

Transcription

Referring again to the structure of the cell, recall that DNA is found in the nucleus (fig. 4.1). Protein synthesis occurs at the ribosomes, which are located in the cytoplasm or on a membrane called the endoplasmic reticulum. Therefore, it is obvious that there must be some intermediary way of getting the DNA message, or code, to the ribosomes. *Messenger RNA* (mRNA) takes the message from the DNA in the nucleus to the ribosome in the cytoplasm.

Just as DNA can serve as a template for the production of itself, it can also serve as a template for the production of mRNA. During this process called transcription, RNA nucleotides complementary to a portion of one DNA strand join. The mRNA that results has a sequence of bases complementary to those of a gene (fig. 4.7). While DNA contains a **triplet code** in which every three bases stand for one amino acid, mRNA contains **codons,** each of which is made up of three bases that also stand for an amino acid (table 4.2). After formation, mRNA moves to the cytoplasm, where ribosomes attach to it.

Translation

The next step leading to protein synthesis is called translation, because the order of the codons in mRNA is translated into a particular order of amino acids in a protein.

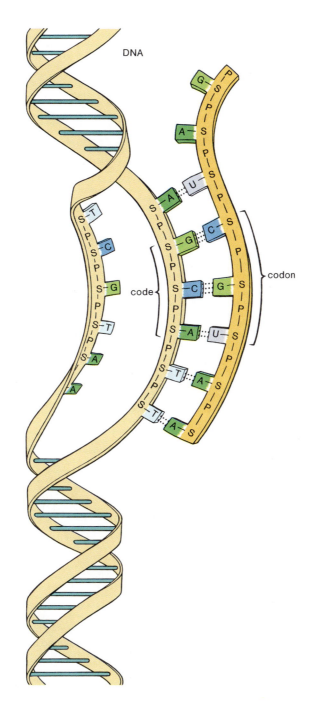

Figure 4.7

Transcription. During transcription mRNA is formed when nucleotides complementary to the sequence of bases in a portion of DNA (*i.e.* a gene) join. Note that DNA contains a code (sequences of 3 bases) and that mRNA contains codons (sequences of 3 complementary bases). *Top*—mRNA transcript is ready to move into the cytoplasm. *Middle*—transcription has occurred and complementary nucleotides have joined together. *Bottom*—rest of DNA molecule.

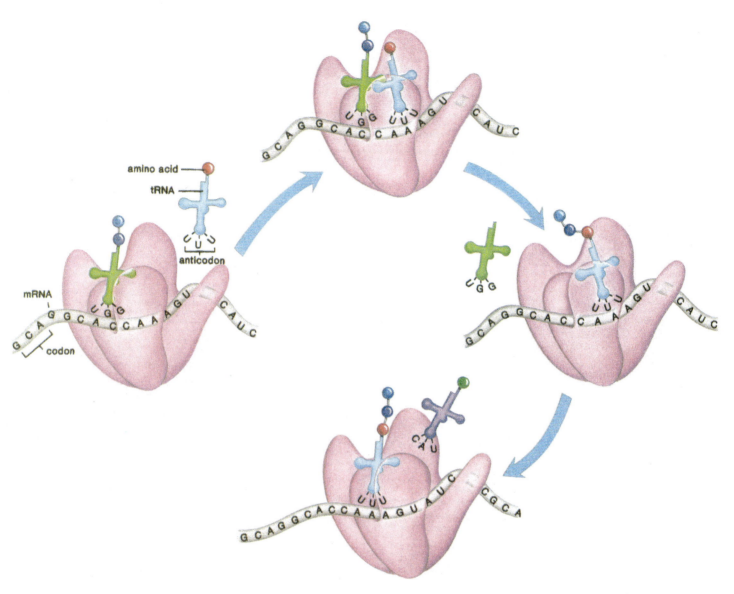

Figure 4.8

During translation, tRNA molecules bring amino acids to the ribosomes in the order dictated by mRNA codons and originally the DNA code. If the codon is ACC, the anticodon is UGG, and the amino acid is the type represented by a purple circle (see table 4.2). Notice that as a tRNA departs the amino acid chain is shifted to a newly arrived tRNA.

For this to happen, free (or unattached) amino acids in the cytoplasm must move to the vicinity of the ribosomes.[2] *Transfer RNA* (tRNA) molecules combine with and bring the amino acids to the ribosomes. There is at least one kind of tRNA for each of the 20 amino acid molecules. A tRNA has three particular nucleotides whose bases make up an **anticodon** (fig. 4.8). Each anticodon is complementary to a particular codon.

2. See footnote 1.

As the ribosomes move along the mRNA, a particular codon becomes prominent. The tRNA, which has the complementary anticodon at one end and the appropriate amino acid at the other, moves into place. Therefore, the sequence of the codons dictates the sequence of the tRNA molecules, and this, in turn, dictates the order of the amino acids in a protein. By this indirect process, the DNA code eventually controls the sequence of amino acids in proteins and thus in enzymes.

The entire transcription-translation sequence is diagrammed in figure 4.9.

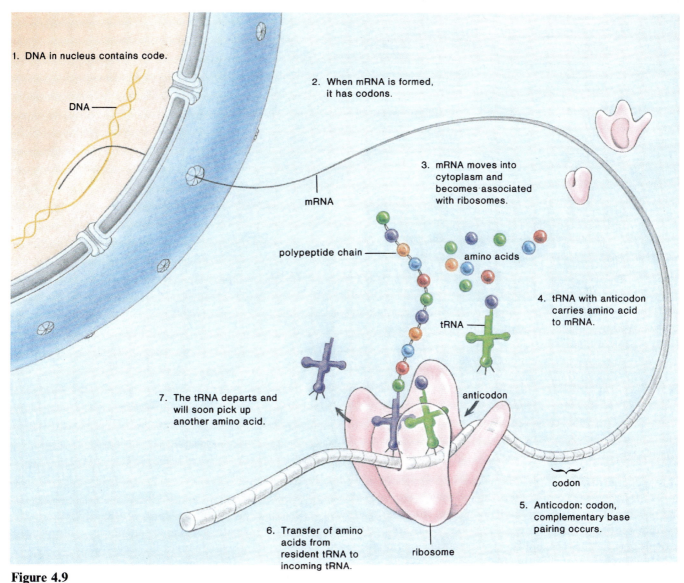

1. DNA in nucleus contains code.

DNA

2. When mRNA is formed, it has codons.

mRNA

3. mRNA moves into cytoplasm and becomes associated with ribosomes.

polypeptide chain

amino acids

4. tRNA with anticodon carries amino acid to mRNA.

tRNA

anticodon

7. The tRNA departs and will soon pick up another amino acid.

codon

5. Anticodon: codon, complementary base pairing occurs.

6. Transfer of amino acids from resident tRNA to incoming tRNA.

ribosome

Figure 4.9

Summary of protein synthesis. Transcription occurs in the nucleus, and translation occurs in the cytoplasm (blue background). During translation, as a ribosome moves along the mRNA, the tRNAs pair with the codons so that the amino acids become sequenced in a particular order.

Steps in Protein Synthesis

It is now possible to list the steps involved in protein synthesis (table 4.3).

1. DNA, which remains in the nucleus, contains a series of bases that serve as a triplet code (a sequence of three bases).
2. One strand of DNA serves as a template for the formation of messenger RNA (mRNA), which contains triplet codons (sequences of three complementary bases).
3. Messenger RNA goes into the cytoplasm and becomes associated with the ribosomes, which are composed of ribosomal RNA (rRNA).
4. Transfer RNA (tRNA) molecules, each of which is bonded to a particular amino acid, have anticodons that pair complementarily to the codons in mRNA.
5. Transfer RNA molecules, along with their amino acids, come to the ribosomes in the order dictated by mRNA, and in this way amino acids become ordered in a particular sequence.

TABLE 4.3 STEPS IN PROTEIN SYNTHESIS

Molecule	Special Significance	Definition
DNA	Code	Sequence of bases in threes
mRNA	Codon	Complementary sequence of bases in threes
tRNA	Anticodon	Sequence of three bases complementary to codon
Amino acids	Building blocks	Transported to ribosomes by tRNAs
Protein	Enzymes and structural proteins	Amino acids joined in a predetermined order

6. As the amino acid chain lengthens, a specific protein begins to form.
7. The transfer RNA molecules repeat the process of transporting amino acids to the ribosomes until the protein molecule is completed.

METABOLIC AND STRUCTURAL DISORDERS

Most genetic disorders are due to the inheritance of a faulty DNA code for either a metabolic protein (enzyme) or a structural protein such as hemoglobin, the respiratory pigment found in red blood cells. **Mutations** are permanent genetic changes, but researchers are beginning to discover ways to treat resulting metabolic defects. For example, SCID children (chapter 3) lack an enzyme that can be administered to them. Research is even going forward to insert a functional gene into their white blood cells (chapter 5).

Germinal mutations arise when an individual inherits an altered sequence of DNA bases. *Somatic mutations* arise when the sequence of bases within somatic cells is altered.

Metabolic Disorders

When DNA contains a code for an inappropriate sequence of amino acids in an enzyme, the enzyme is nonfunctional. For example, one particular metabolic pathway in cells is as follows:

$$A \text{ (phenylalanine)} \xrightarrow{E_A} B \text{ (tyrosine)} \xrightarrow{E_B} C \text{ (melanin)}$$

If a faulty code for enzyme E_A is inherited, a person is unable to convert the molecule A to B. Phenylalanine

Figure 4.10
Albino individuals lack melanin (skin pigment) because they inherited alleles that code improperly for an enzyme.

builds up in the system and the excess causes mental retardation and the other symptoms of PKU (chapter 2).

In this same pathway, if a person inherits a faulty code for enzyme E_B, then C cannot be converted to B and the individual is an albino (fig. 4.10).

Structural Disorders

Since DNA codes for all proteins, genetic defects can involve proteins that are not enzymes. For example, sickle-cell disease is caused by the inheritance of a faulty code for hemoglobin. It is now known that there is only one amino acid change between normal hemoglobin and sickle-cell hemoglobin (fig. 4.11). This one amino acid change causes the hemoglobin molecules to adhere to one another in such a way that the cells take on a sickle shape. One author explains the situation in the following way:

The symptoms of sickle-cell disease occur because the cells die off quickly and because the sickle-shaped cells clog the blood vessels. The cells have the sickle shape because of the rodlike clumping of molecules within the cells. The molecules clump that way because in one tiny spot a chemical attaches on differently than it does in normal red

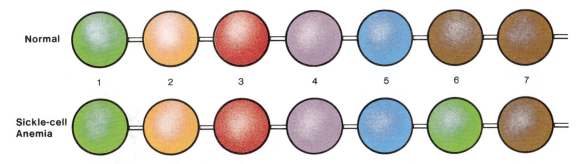

Figure 4.11

Sickle-cell disease is caused by the inheritance of a faulty code for hemoglobin. A mutation has occurred so that DNA codes for the wrong amino acid at number 6. The normal sequence of amino acids (represented by colored balls) is shown above the sequence in sickle-cell hemoglobin.

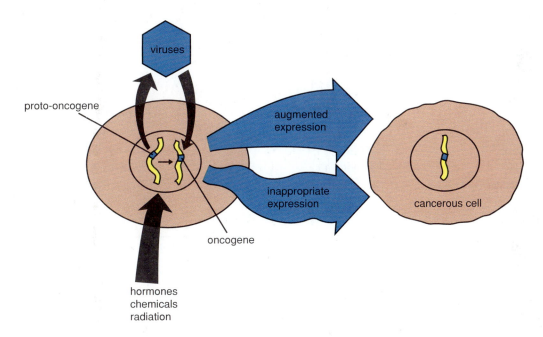

Figure 4.12

Cancer develops when an oncogene is present in a cell. A virus can pass an oncogene to a cell; or a proto-oncogene can become an oncogene due to a mutation caused by a chemical or radiation. Compared to the normal gene, the oncogene either expresses itself to a greater degree than normal or else expresses itself inappropriately.

blood cell molecules. The tiny chemical (an amino acid) attaches that way because of the genes that were inherited from the person's ancestors.[3]

Cancer is characterized by irregular and uncontrolled growth of cells. For a time, the cancer remains at the site of origin, but eventually cancer cells invade underlying tissues, they become detached, and they are carried by the lymphatic and circulatory systems to other parts of the body, where new cancer growth may begin.

It has long been believed that cancer begins with a change in the DNA, but the exact nature of the change was unknown. Investigators have determined that cells contain genes called proto-oncogenes (proto = before; onco = tumor), which become **oncogenes**, that is, cancer-causing genes (fig. 4.12). These genes are not alien to the cell; they are normal, essential genes that have undergone a mutation. An environmental mutagen (any factor that increases the chances of a mutation), such as various chemicals and radiations, can cause cancer when they bring about the conversion of a proto-oncogene to an oncogene. Also, cancer-causing viruses are believed to pick up a proto-oncogene from one person and bring it into a cell of another person, where it is an active oncogene. Other factors can also contribute to the development of cancer. ∎

3. S. M. Linde, *Sickle Cell* (New York: Pavilion Publishing, 1972).

SUMMARY

Chromosomes are composed of DNA, a double-stranded molecule that contains a triplet code utilizing four bases: A, T, C, G. By means of complementary base pairing, DNA serves as a template for its own replication and for the transcription of messenger RNA, a single-stranded nucleic acid that contains codons utilizing the four bases A, U, C, G.

Messenger RNA moves from the nucleus to the cytoplasm, where ribosomes composed of ribosomal RNA aid in the process of translation. Transfer RNA molecules carry amino acid molecules to the ribosomes, where they become linked together in the order predetermined by the DNA code. The resulting protein molecules have various functions in cells but are notably

enzymes involved in a metabolic pathway.

Genetic diseases are often inborn errors of metabolism because the inheritance of a faulty genetic code leads to a defective enzyme. Sickle-cell disease, on the other hand, is caused by the inheritance of a faulty code for hemoglobin, a component of red cells.

KEY TERMS

adenine (A) 43

amino acids 48

anticodon 50

cancer 53

codons 48

complementary base 43

cytosine (C) 43

DNA (deoxyribonucleic acid) 43

double helix 43

enzymes 48

guanine (G) 43

messenger RNA (mRNA) 46

mutations 52

oncogenes 53

proteins 48

replication 45

ribosomal RNA (rRNA) 46

ribosomes 43

RNA (ribonucleic acid) 43

thymine (T) 43

transcription 48

transfer RNA (tRNA) 46

translation 48

triplet code 48

REVIEW QUESTIONS

1. Describe the structure of DNA and explain how DNA duplicates itself.

2. How does DNA structure differ from RNA structure?

3. Define nucleotide, nucleic acid, base, sugar, and phosphate.

4. Describe the structure of proteins. How many different types of amino acids are there?

5. Compare code to codon to anticodon. What does the term

complementary base pairing mean?

6. If ATCGTACCG were in DNA, what would the code be? the codons? the anticodons?

7. Where are *mRNA, rRNA,* and *tRNA* located in the cell?

8. Explain protein synthesis, starting with DNA and finishing with the protein. Be sure to

mention code, codon, anticodon, mRNA, tRNA, amino acids, and ribosomes in your answer.

9. What is wrong with someone who has PKU? Describe in terms of faulty code and faulty enzyme.

10. What is wrong with someone who has sickle-cell disease? Describe in terms of faulty code and faulty hemoglobin.

CRITICAL THINKING QUESTIONS

1. A card file stores information and so does DNA. Comparing these two, the piece of furniture that holds the card file; the cards; and the letters on the cards are equivalent to what in

the organism? DNA stores information for what?

2. Why is replication of DNA necessary; that is, why does every cell need a copy of DNA?

3. Considering that there are many different types of living things, why is it beneficial that DNA can mutate?

C H A P T E R

Biotechnology

Biotechnology, the use of a natural biological system to produce a product or to achieve an end desired by human beings, is not new. Plants and animals have been bred to yield a particular phenotype since the dawn of civilization. In addition, the biochemical capabilities of microorganisms have been exploited for a very long time. For example, the baking of bread and the production of wine are dependent on yeast cells to carry out fermentation reactions.

Today, however, biotechnology is first and foremost an industrial process in which we are able genetically to engineer bacteria to produce human proteins or other protein products of interest (fig. 5.1). Biotechnology is also being investigated as a way to alter the genotype and subsequently the phenotype of animals and plants. It is also hoped that eventually it may be possible to carry out gene therapy in human beings to correct inherited disorders.

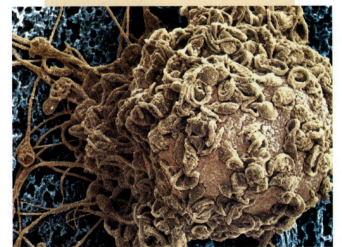

Fertilization is occurring.

BIOTECHNOLOGY TECHNIQUES

Bioengineered, or genetically engineered, bacteria contain a foreign gene, such as a human gene, and therefore are capable of producing a human protein (fig. 5.2). Often a human gene is carried into a bacterium as a part of recombinant DNA.

Recombinant DNA

Recombinant DNA contains DNA from two or more different sources, such as bacterial and human DNA. Plasmids are frequently used to carry recombinant DNA into bacteria. This is logical because plasmids are taken from bacteria in the first place. A bacterial cell is different from a human cell—it does not have a nucleus, but it does have a single chromosome and often a **plasmid,** a small accessory ring of DNA outside the chromosome.

The introduction of a human gene into a plasmid to produce recombinant DNA requires two different types of enzymes. A *restriction enzyme* is used to cut open bacterial plasmid DNA and to slice out the human gene from human DNA (fig. 5.2). *DNA ligase enzyme* is used to seal the cut ends after the human gene is inserted into the plasmid. Bacteria will take up recombined plasmids. There, the plasmid will be copied whenever the bacterium reproduces. Eventually, there are many copies of the plasmid and, therefore, many copies of the human gene, such as the gene that codes for insulin, a protein that helps regulate sugar metabolism in humans. People who have diabetes sometimes need an injection of insulin every day to control their condition.

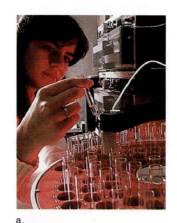

a.

b.

c.   d.

Figure 5.1

Biotechnology is an industrial endeavor. *a.* Laboratory procedures are adapted to mass-produce a biotechnology product. *b.* Microbes that produce the product are grown in huge tanks called fermenters. *c.* The product is purified and (*d*) packaged.

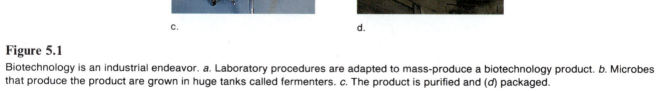

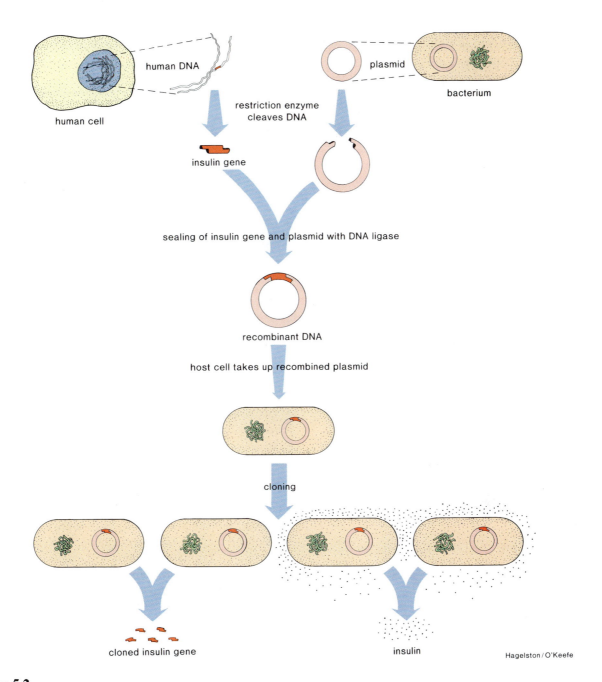

human cell

human DNA

plasmid

bacterium

restriction enzyme
cleaves DNA

insulin gene

sealing of insulin gene and plasmid with DNA ligase

recombinant DNA

host cell takes up recombined plasmid

cloning

cloned insulin gene

insulin

Hagelston / O'Keefe

Figure 5.2

Methodology for cloning a human gene (e.g., insulin gene). Both human DNA and a bacterial plasmid (circular ring of DNA) are cleaved by a restriction enzyme in such a way that it is possible to insert the human DNA into the plasmid. The enzyme called DNA ligase seals the recombinant plasmid which is taken up by a new bacterium. When this bacterium reproduces, the human gene is cloned. When the human gene produces insulin, a biotechnology product is available. The investigator can retrieve either the cloned gene or the product for analysis or use.

When the recombinant DNA technique was first devised, scientists were concerned about creating mutant bacteria that could possibly escape into the environment and cause harm. However, nothing untoward has yet developed because of DNA-splicing techniques. Indeed, a number of benefits have resulted. Still, concern over potential hazards persists, and the governmental approval process for genetically engineered organisms and their products can take years. Do you support a cautious and restrictive approach or a free and open approach? What type of approval process would you suggest? by whom?

Other Biotechnology Techniques

Aside from recombinant DNA, other techniques are common in biotechnology, including the following:

1. Gene sequencing, or determining the order of nucleotides in a gene. There are automated DNA sequencers that make use of computers to sequence genes at a fairly rapid rate.
2. Manufacture of a gene by using a DNA synthesizer. The nucleotides are joined in the correct sequence, or if desired, a mutated gene is prepared by altering the sequence.
3. Insertion of a gene directly into a host cell without using a plasmid. Animal cells, in particular, do not take up plasmids, but DNA can be microinjected into them.

BIOTECHNOLOGY PRODUCTS

Table 5.1 shows some of the types of biotechnology products available.

Hormones and Similar Types of Proteins

One impressive advantage of biotechnology is it allows mass production of proteins that are very difficult to obtain otherwise. For example, insulin was once extracted from the pancreas glands of slaughtered cattle and pigs; it was expensive and sometimes caused allergic reactions in recipients. Previously human growth hormone was extracted from the pituitary gland of cadavers, and it took 50 glands to obtain enough for one dose. Few of us knew of tPA (tissue plasminogen activator), a protein present in the body in minute amounts that activates an enzyme to dissolve blood clots. Now tPA is a biotechnology product used to treat heart attack victims. Erythropoietin also was not available, but now it can be used to prevent anemia in patients with kidney failure receiving hemodialysis, for example.

TABLE 5.1 REPRESENTATIVE BIOTECHNOLOGY PRODUCTS
Hormones and Similar Types of Proteins—Treatment of Humans
Insulin
Growth hormone
tPA (tissue plasminogen activator)
Erythropoietin
Clotting factor VIII*
Human lung surfactant*
Atrial natriuretic factor*
DNA Probes—Diagnostics in Humans
Various diseases (e.g., sexually transmitted diseases)
Inherited disorders (e.g., cystic fibrosis)
Various cancers (e.g., chronic myelogenous leukemia)
Detection of criminals
Paternity suits
Vaccines—Use in Humans
Hepatitis B
AIDS*
Herpes* (oral and genital)
Hepatitis A* and C*
Malaria*

*Expected in the near future. These may be available now, but presently they are not made by recombinant DNA technology.

Other biotechnology products that are expected soon are clotting factor VIII to treat hemophilia; human lung surfactant for premature infants with respiratory distress syndrome; and atrial natriuretic factor to treat hypertension. Factor VIII and lung surfactant have been available from other sources but not as a biotechnology product.

Hormones for use in animals are also biotechnology products. Cows given bovine growth hormone (bGH) produce 25% more milk than usual, which should make it possible for dairy farmers to maintain fewer cows and to cut down on overhead expenses.

DNA Probes

A **DNA probe** is a piece of single-stranded DNA, often radioactive, that can be used to detect the presence of a certain allele in a cell or body fluid. Today a probe can be prepared by using a DNA synthesizer that links nucleotides together in same order as a portion of a gene. The probe will bind by complementary base pairing to a particular DNA sequence. In the process called DNA fingerprinting, several probes are used to compare the DNA in hair or body fluid taken from the scene of a crime with

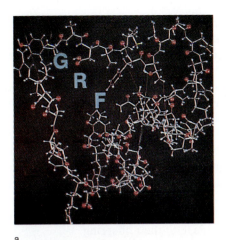

a.

b.

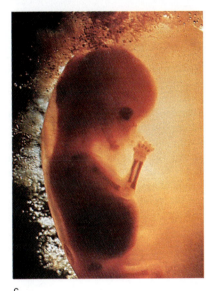

c.

Figure 5.3

Possible biotechnology scenario. *a.* A protein is removed from a cell, and the amino acid sequence is determined. In this photo, the hormone is growth hormone releasing factor (GRF). From this, the sequence of nucleotides in DNA can be determined. *b.* The DNA synthesizer can be used to string nucleotides together in the correct order. *c.* A small section of the gene can be used as a DNA probe to test fetal cells for the presence of the gene. The manufactured gene could also be placed in a bacterium to produce more of the protein, which then could be used as treatment if needed.

that of DNA obtained from blood cells of a suspected criminal. If the data match, the criminal is caught just as if he or she had left fingerprints behind. DNA probes can also be used in court cases to identify the actual parents, either mother or father, of an individual.

In medicine, DNA probes are used to diagnose an infection by indicating if the gene of an infectious organism is present. In most instances, blood tests are used to detect the presence of antibodies to a pathogen rather than the pathogen itself. For example, the blood test for AIDS does not detect the AIDS virus itself, but instead it confirms the presence of antibodies to the AIDS virus in the blood.

The use of a probe for the virus allows the detection of an infection even before antibodies are present; therefore, earlier treatment is possible.

When available, a DNA probe can tell us whether a gene coding for a hereditary defect might be present (fig. 5.3). When a couple is using in vitro fertilization a probe can be used to test for a genetic disorder even before an embryo is placed in the uterus. This ensures that a couple will only have children that are free of a particular genetic disorder.

DNA probes are also used to help investigators map the human chromosomes. This is a lengthy process that will take many years to complete. More and more uses for

DNA probes are being found because the polymerase chain reaction allows "targeted" pieces of DNA to be amplified before the probe is applied. The reading entitled "Polymerase Chain Reaction" gives more information about this process.

Vaccines

Vaccines are used to make people immune to an infectious organism. People who receive a vaccine do not become ill when exposed because a vaccine causes the buildup of antibodies that protect them from the infectious organism. In the past vaccines were made from treated bacteria or viruses. But bacteria and viruses have surface proteins, and a gene for just one of these can be placed in a plasmid. When bioengineered bacteria produce many copies of the surface protein, these copies can be used as a vaccine. Right now, the only vaccine produced through biotechnology is hepatitis B, but others are expected, such as a vaccine for malaria and another for AIDS.

TRANSGENIC ORGANISMS

Free-living organisms in the environment that have had a foreign gene inserted into them are called **transgenic organisms.** Transgenic bacteria are an aid to farmers when they protect plants from pests and frost. Soon, there may be transgenic plants that are themselves resistant to pests, that require less fertilizer and herbicides than usual, or that are able to grow under unfavorable environmental conditions. Our interest centers mostly on transgenic animals, however.

Genetic engineering of animals has begun. Animal cells will not take up plasmids, but it is possible to microinject foreign genes into eggs before they are fertilized. The most common procedure attempted today is to mi-

croinject bovine growth hormone (bGH) into the eggs of various animals. It is hoped that the gene will establish itself and be transmitted to all the cells of the developing organism and even be passed along to the next generation of offspring. This procedure has been used in fishes, chickens, cows, pigs, rabbits, and sheep in the hope of producing bigger varieties.

In another experiment, the sheep gene that codes for the milk protein betalactoglobulin (BLG) was microinjected into mice eggs. (Mice normally produce a milk that has no BLG at all.) Of 46 offspring successfully weaned, 16 carried the BLG sequence, and the females among them later produced BLG-rich milk. Some of these females passed on the BLG gene to their offspring.

GENE THERAPY IN HUMANS

Investigators are striving to use biotechnology as a means of curing or treating human genetic disorders and treating other human ills as well.

Bioengineering Bone Marrow Stem Cells

Some genetic disorders are caused by the inheritance of a faulty code for a particular protein that functions in blood cells. Red bone marrow contains stem cells that give rise to the many types of blood cells in the body. Some investigators believe that it will be possible to use a virus to introduce a normal functioning gene into stem cells, and thereafter all types of blood cells will produce functioning enzymes. The virus to be used is a *retrovirus,* which uses RNA instead of DNA for its genes.

All viruses are composed of just two parts: a coat of protein and a nucleic acid core. Only the nucleic acid enters a host cell. Most viruses then reproduce themselves by producing more nucleic acid enclosed by protein. Retro-

The polymerase chain reaction (PCR) is a way to make multiple copies of a single gene, or any specific piece of DNA, in a test tube. The process is very specific—the targeted DNA sequence can be less than one part in a million of the total DNA sample! This means then that a single gene among all the human genes can be amplified (copied) using PCR.

The process takes its name from DNA polymerase, the enzyme that carries out DNA replication in a cell. The reaction is a chain reaction because DNA polymerase is allowed to carry out replication over and over again until there is a million or more copies of the targeted DNA. The polymerase chain reaction will not replace DNA cloning. Cloning provides many more copies of gene than this and still will be used whenever a large quantity of a gene or a protein product is needed.

Before carrying out PCR, it is necessary to have available *primers,* sequences of about 20 bases that are complementary to the bases on either side of the "target DNA." The primers are needed because DNA polymerase does not start the replication process; it continues or extends the process. After the primers bind complementarily to the DNA strand, DNA polymerase copies the target DNA and only the target DNA between the primers. Therefore, PCR is very specific.

PCR has been in use for several years, but a recent advance has been the intro-duction of automated PCR machines, allowing most any laboratory to carry out the procedure. Automation became possible after a thermostable DNA polymerase was extracted from a bacterium. Using this enzyme means that there is no need to add more DNA polymerase each time a high temperature is used to separate double-stranded DNA so that replication can occur once again.

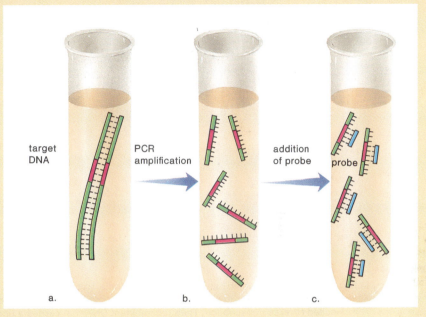

Figure 5.A

PCR analysis. *a.* DNA is removed from a cell and placed in a test tube along with appropriate primers, DNA polymerase, and a supply of nucleotides. *b.* Following PCR amplification, many copies of target DNA (red) are present. *c.* Binding of a labeled DNA probe (blue) allows the scientist to determine that a particular DNA segment was indeed present in the original sample.

Figure 5.A shows that after PCR amplification it becomes a lot easier to use a probe to detect that target DNA is present. (The binding of the probe is detectable because the probe is labeled either radioactively or with a fluorescent dye.) Therefore, because of the polymerase chain reaction DNA probes are increasingly used for many purposes.

viruses are unique because they have an enzyme that on occasion transcribes RNA to DNA, which then becomes part of the host cell's DNA. When RNA viruses are used for gene therapy, they have been equipped with recom-binant RNA (fig. 5.4). After the recombinant RNA enters a human cell, such as a bone marrow stem cell, *reverse transcription* occurs and recombinant DNA enters a human chromosome.

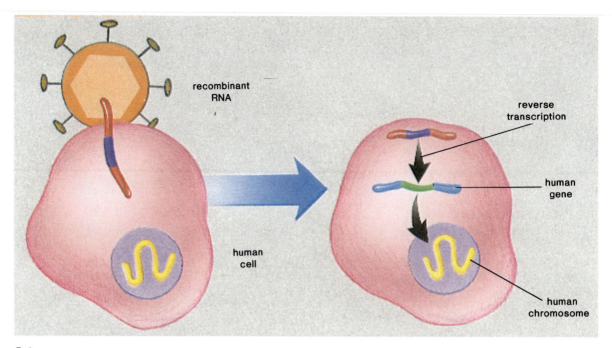

Figure 5.4

Gene therapy in human. A retrovirus, which has an RNA chromosome, will be used for gene therapy in humans. First the retrovirus must be equipped with recombinant RNA. After the recombinant RNA enters the human cell, it is changed to recombinant DNA by a special viral enzyme. This is called reverse transcription. Then recombinant DNA that carries a normal human gene will enter the human chromosome. The individual is cured of a genetic disorder when the normal gene directs the production of its product.

Bioengineering Other Types of Cells

Using the same method described in figure 5.4, a retrovirus can also be used to bioengineer human lymphocytes (a type of white blood cell) directly. It is expected that soon the gene that codes for an enzyme needed by patients with SCID (the "bubble-baby" disease, chapter 3) will be introduced into their lymphocytes and thereafter these children will be able to lead a normal life.

Cells that normally line blood vessels (endothelial cells) are easily obtained and cultured in the laboratory, and they can be bioengineered to make a particular enzyme. Whereas bone marrow stem cells and lymphocytes can be withdrawn and return to the body by injection, endothelial cells need a special delivery system. One novel idea is to put the cells in *organoids,* artificial organs that can be implanted in the abdominal cavity. Organoids have been made by coating angel-hair Gore-Tex fibers with a substance called collagen and adding a growth factor for blood vessels (fig. 5.5). Once in the body, the endothelial cells begin to line the newly developing blood vessels, which spread out to adjoining organs.

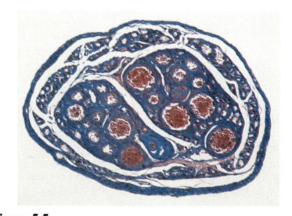

Figure 5.5

This organoid was made by treating Gore-Tex fibers with a growth factor that stimulates blood vessel formation. The organoid is designed to carry bioengineered cells that will carry a normal gene to make up for a defective gene inherited by the individual.

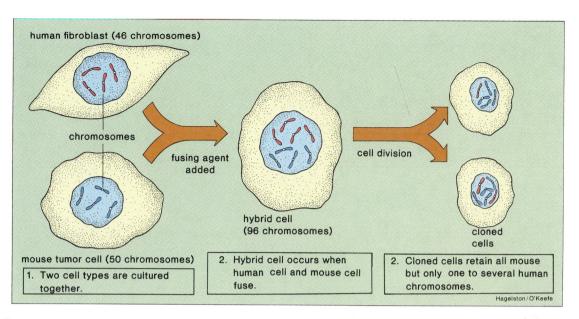

Figure 5.6

In the presence of a fusing agent, human fibroblast cells sometimes join with mouse tumor cells to give hybrid cells with nuclei that contain both types of chromosomes. Subsequent cell division of the hybrid cell produces clones that have lost most of their human chromosomes, allowing the investigator to study these chromosomes and their products separate from all other human chromosomes.

MAPPING THE HUMAN CHROMOSOMES

If investigators knew the order and precise location of the genes on the human chromosomes, it would facilitate laboratory research and medical diagnosis and treatment. Several methods have been used to attempt mapping the human chromosomes, and we will discuss some of these.

Human-Mouse Cell Data

When human and mouse cells are mixed together in a laboratory dish, in the presence of a fusing agent, they will fuse (fig. 5.6). As the cells grow and divide, some of the human chromosomes are lost, and eventually the daughter cells contain only a few human chromosomes, each of which can be recognized by its distinctive banding pattern (fig. 1.2). Analysis of the proteins made by the various human-mouse cells enables scientists to determine which genes are to be associated with which human chromosome.

Sometimes it is possible to obtain a human-mouse cell that contains only one human chromosome or even a portion of a chromosome. This technique has been very helpful to those researchers who have been studying the genes located on the number-21 chromosome.

Genetic Marker Data

A **genetic marker** is an observed DNA difference between individuals that sometimes can be associated with a genetic disorder. Genetic markers are discovered by using restriction endonuclease enzymes, because each one always cleaves DNA at a specific sequence of bases. For example, there is a restriction enzyme that always cleaves double-stranded DNA when it has this sequence of bases:

$$. . . . . G\,A\,A\,T\,T\,C . . . . .$$
$$. . . . . C\,T\,T\,A\,A\,G . . . . .$$

If an individual has this sequence of bases at a particular location, then the restriction enzyme is able to cleave the chromosome at this location. If an individual lacks this sequence of bases, then the restriction enzyme is unable to cleave the chromosome at this location. Scientists say that "restriction fragment length polymorphisms (RFLPs)" can be observed. Polymorphisms mean many changes in structure; in this case polymorphisms exist in the fragment lengths following restriction enzyme digest.

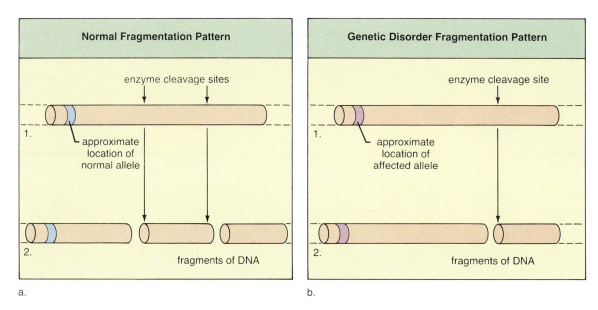

Figure 5.7

Use of a genetic marker to detect the approximate location of an allele or to test for a genetic disorder when the exact location of the allele is unknown. *a.* DNA from the normal individual has certain restriction enzyme cleavage sites near the allele in question. *b.* DNA from another individual lacks one of the cleavage sites, and this loss indicates that they almost certainly have the genetic disorder because experience has shown that this genetic marker is almost always present when an individual has the disorder. In other cases, a gain in a cleavage site is the genetic marker.

Genes can be assigned a location on a chromosome according to their relative relationship to genetic markers. Genetic markers can also be used to tell if one has a genetic disorder when a particular marker is always inherited with a particular genetic disorder. The test for the genetic disorder consists of testing an individual for the presence of the marker, instead of for the specific faulty allele (fig. 5.7). For a marker to be dependable it should be inherited with the defective allele at least 98% of the time.

The current tests for sickle-cell disease, Huntington disease, and Duchenne muscular dystrophy are all based on the presence of a marker.

Human Genome Project

The goal of the Human Genome Project is to identify the location of the approximately 100,000 human genes on all the chromosomes. In order to create this *genetic map,* ge-

netic markers will be used to index the chromosomes. Known and newly discovered genes will be assigned locations between the markers.

The project also wants to eventually know the sequence of the 3 billion bases in the human genome. In order to create this *physical map* researchers will use laboratory procedures to determine the sequence of the DNA bases.

Recently it's been discovered that genetic markers contain unique stretches of DNA called sequence-tagged sites, or STSs. Hopefully, these can be used to create links between the genetic map and the physical map.

The human genome project will require millions of dollars and many years to complete. Just how successful and useful it will be cannot yet be determined. ∎

SUMMARY

Biotechnology is being expanded greatly by DNA technology. To achieve a genetically engineered bacterium, often plasmids are used to carry a human gene into a bacterium. First, restriction endonuclease enzymes are used to cleave plasmid DNA and human DNA. Second, after the human gene has been inserted in the plasmid, DNA ligase is used to seal the cut ends. After a bacterium takes up a plasmid and reproduces, the recombinant plasmid is copied, and therefore is cloned. The human protein produced by the bioengineered bacterium can be collected, purified, and sold. The types of biotechnology products available today are hormones, DNA probes, and vaccines.

Transgenic organisms also have been made. Bacteria can be engineered to increase the health of plants. Also, plants that require less fertilizer and are resistant to pests and herbicides are being developed. The possibility also exists of developing improved livestock through biotechnology.

Human gene therapy is being investigated. Researchers are interested in engineering bone marrow stem cells, lymphocytes, and endothelial cells. The United States government is committed to mapping and sequencing the entire human genome. It is hoped this information will eventually assist human gene therapy.

KEY TERMS

bioengineered 55
biotechnology 55
DNA probe 58

genetic marker 63
plasmid 55

recombinant DNA 55
transgenic organisms 60

REVIEW QUESTIONS

1. Describe and explain a methodology for cloning a gene.

2. List and explain other types of laboratory procedures used in biotechnology.

3. Categorize and give examples of types of biotechnology products available today.

4. What is a DNA probe? When do scientists use DNA probes?

5. What is the polymerase chain reaction (PCR)? How does PCR facilitate the use of probes?

6. Describe the current situation in regard to transgenic animals.

7. Describe the methodology for bioengineering human cells using a retrovirus. How might bioengineered endothelial cells be introduced into the body?

8. What methods are available for mapping the human chromosomes?

CRITICAL THINKING QUESTIONS

1. Why would you expect insulin produced by bioengineered bacteria to have the same structure and function in the human body as insulin produced by a human being?

2. It is possible to produce a transgenic plant that contains a functioning animal gene. What does this tell you about plant cells?

FURTHER READINGS FOR PART I

Antebi, E., and D. Fishlock. 1986. *Biotechnology: Strategies for life.* Cambridge, Mass.: The MIT Press.

Berns, M. W. 1983. *Cells,* 3d ed. New York: Holt, Rinehart & Winston.

Bishop, J. M. March 1982. Oncogenes. *Scientific American.*

Dickerson, R. E. December 1983. The DNA helix and how it is read. *Scientific American.*

Drlica, K. 1984. *Understanding DNA and gene cloning.* New York: John Wiley & Sons.

Elkington, J. 1985. *The gene factory: Inside the science and business of biotechnology.* New York: Carroll and Graf Publishers.

Feldman, M., and L. Eisenback. November 1988. What makes a tumor cell metastatic? *Scientific American.*

Gilbert, W., and L. Villa-Komaroff. April 1980. Useful proteins from recombinant bacteria. *Scientific American.*

Glover, D. M. 1984. *Gene cloning: The mechanics of DNA manipulation.* New York: Chapman and Hall.

Kieffer, G. H. 1987. *Biotechnology, genetic engineering and society.* Reston, Va.: National Association of Biology Teachers.

Lake, J. A. July 1981. The ribosome. *Scientific American.*

Lerner, I. M., and W. Libby. 1976. *Heredity, evolution, and society,* 2d ed. San Francisco: W. H. Freeman & Co.

McIntosh, J. Richard and McDonald, Kent L. October 1989. The mitotic spindle. *Scientific American.*

Mullis, Kary B. April 1990. The unusual origin of the polymerase chain reaction. *Scientific American.*

Nathans, Jeremy. February 1989. The genes for color vision. *Scientific American.*

Pardee, A. B., and G. Prem veer Reddy. *Cancer fundamental ideas.* Burlington, N. C.: Carolina Biological Supply Co.

Patterson, D. August 1987. The causes of Down syndrome. *Scientific American.*

Radman, M., and R. Wagner. August 1988. The high fidelity of DNA duplication. *Scientific American.*

Ross, Jeffrey. April 1989. The turnover of messenger RNA. *Scientific American.*

Ruddle, F. H., and R. S. Jucherlapati. July 1974. Hybrid cells and human genes. *Scientific American.*

Scientific American. October 1985. The molecules of life. Entire issue.

Smith, D. W., and A. A. Wilson. 1973. *The child with Down's syndrome.* Philadelphia: W. B. Saunders.

Stent, G. S. 1978. *Molecular genetics: An introductory narrative,* 2d ed. San Francisco: W. H. Freeman & Co.

Sutton, H. E. 1988. *An introduction to human genetics,* 2d ed. New York: Holt, Rinehart & Winston.

Tompkins, J. S., and C. Rieser. June 1986. Special report: Biotechnology. *Science Digest.*

Verma, Inder M. November 1990. Gene therapy. *Scientific American.*

White, R., and J. Lalouel. February 1988. Chromosome mapping with DNA markers. *Scientific American.*

Winchester, A. M. 1983. *Human genetics,* 4th ed. Columbus, Ohio: Charles E. Merrill Publishing.

II

Human Reproduction

Human beings have two sexes, male and female. The anatomy of each sex functions to produce sex cells, which join prior to the development of a new individual. Humans develop in the uterus of the female. The steps of human development can be outlined from the fertilized egg to the birth of a child.

We are in the midst of a sexual revolution, and we have the freedom to experience varied lifestyles. With freedom comes a responsibility to be familiar with the biology of reproduction and potential health hazards, not only for ourselves but also for future children. ■

Sex Hormones and Secondary Sex Characteristics

The human species is dimorphic—its members exist in two forms, male and female. The differences in appearance observed in figure 6.1 are brought about and maintained by the sex hormones.

SEX HORMONES

Both male and female sex organs develop from similar embryonic tissues in the abdominal cavity. The sex organs are the **primary sex characteristics** of humans. Specific genes on the Y chromosome cause the development of testes (chapter 3). Once testes are present, they begin to produce the male sex hormones. These are not only responsible for the development of sex organs, they are also responsible for the maturation of the male organs and the appearance of male **secondary sex characteristics** such as facial hair, greater musculature, and a deepened voice, all of which do not appear until puberty (fig. 6.1a). **Puberty** occurs between ages 9 and 14 and is that time in life when secondary sex characteristics develop.

Genes on the X chromosome in the absence of a Y chromosome lead to embryonic development of ovaries. The female sex hormones produced by the ovaries promote the maturation of the female sex organs (primary sex characteristics) plus the appearance of the secondary sex characteristics such as breast development and widening hips at puberty (fig. 6.1b).

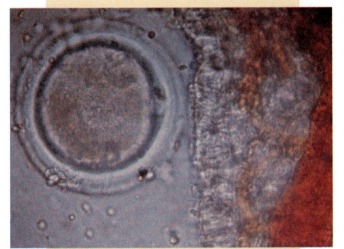

The zygote, a fertilized egg.

Cellular Activity of Sex Hormones

Sex hormones act at the cellular level and are believed to enter only those cells that have specific receptors to receive them (fig. 6.2). Once the hormone has combined with the receptor, the complex moves into the nucleus, where it binds with a portion of DNA. Thereafter, there is accelerated synthesis of messenger RNA with subsequent protein synthesis. It is believed that each sex hormone promotes the activity of certain genes, and it is this action that assists the development of maleness and femaleness.

Sex hormones, like other hormones, reach their target organs by way of the bloodstream. A **hormone** may be defined as a chemical produced by one set of cells but that affects a different set of cells. Hormones are produced by glands called **endocrine glands,** or glands of internal secretion, because their products are distributed in the bloodstream.

Endocrine Glands of Males and Females

The major endocrine glands of the body are shown in figure 6.3, and table 6.1 lists them for easy reference. The glands of special interest to us are the pituitary gland, the adrenal glands, and the gonads (testes in males and ovaries in females).

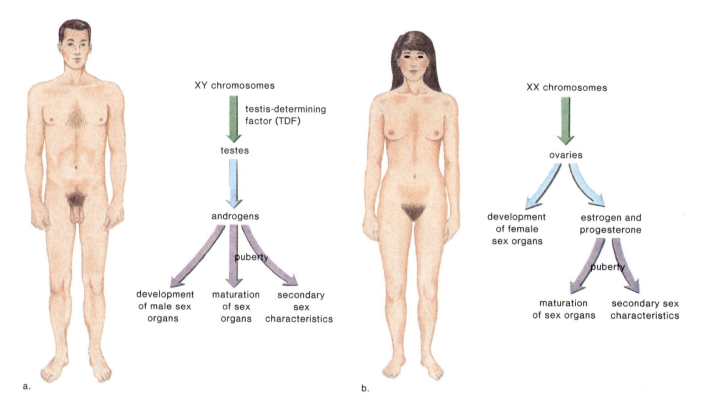

Figure 6.1

Sex hormones and secondary sex characteristics. *a.* It is proposed that a Y chromosome carries a gene called the testis-determining factor which causes a fetus to develop testes. The testes produce androgens, the male sex hormones and these hormones cause the development of other male sex organs. At the time of puberty androgens cause the maturation of sex organs and the appearance of male secondary sex characteristics. *b.* The X chromosome has no testis-determining factor and therefore the fetus develops ovaries instead of testes and also the other female sex organs. At the time of puberty the ovaries produce the female sex hormones, estrogen and progesterone, which cause the maturation of sex organs and the appearance of female secondary sex characteristics.

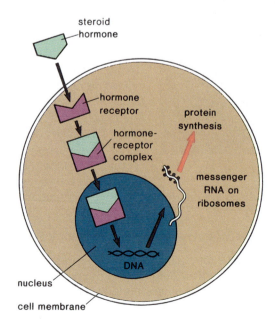

Figure 6.2

Metabolic action of sex hormones. Sex hormones are all steroid (a type of molecule) hormones. Only certain cells that have specific receptors to receive sex hormones are affected by them. When the hormone combines with the receptor, the complex moves into the nucleus, where it activates a certain gene, leading to transcription and translation of a particular enzyme which brings about a metabolic effect and changes the cell in some way.

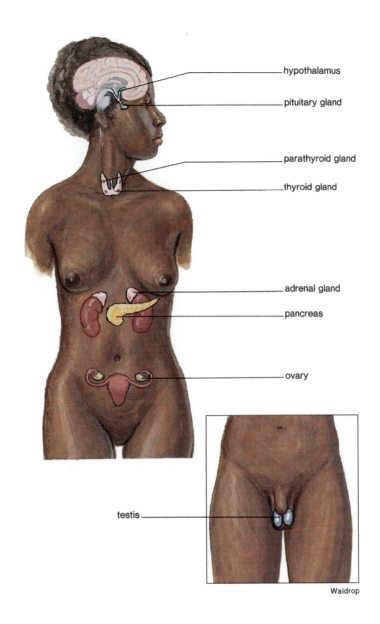

hypothalamus

pituitary gland

parathyroid gland

thyroid gland

adrenal gland

pancreas

ovary

testis

Waldrop

Figure 6.3

The location of certain endocrine (hormone-secreting) glands in the body. The hypothalamus, a part of the brain, controls the secretions of the pituitary gland which lies at the base of the brain. The pituitary gland is divided into two parts, the anterior and posterior pituitary. The anterior pituitary controls the secretions of the thyroid gland, the adrenal cortex, and the sex organs. The adrenal cortex (the outer part of the adrenal gland) and the sex organs (ovaries in females and testes in males) produce sex hormones which promote maturation of sex organs and promote and maintain secondary sex characteristics.

TABLE 6.1   MAJOR ENDOCRINE GLANDS AND THEIR MAJOR HORMONES

Gland	Hormone	Function
Pituitary gland		
Anterior	Thyrotropic-stimulating (TSH)	Stimulates thyroid
	Growth (GH) (somatotropic)	Stimulates growth
	Andrenocorticotropic (ACTH)	Stimulates adrenal cortex
	Prolactin (lactotropic)	Stimulates milk production after delivery
	Gonadotropic Follicle-stimulating (FSH) Luteinizing (LH)	Stimulates gonads
Posterior	Antidiuretic (ADH) (vasopressin)	Regulates water reabsorption by kidneys
	Oxytocin	Stimulates milk letdown
Adrenal gland		
Cortex	Cortisol	Involved in sugar metabolism and relieves stress
	Aldosterone	Regulates salt reabsorption by kidneys
	Sex hormones	See androgens and estrogen below
Medulla	Adrenalin	Fight or flight
Pancreas	Insulin	Lowers blood sugar
Gonads		
Testes	Androgens (testosterone) }	{ Promote maturation of sex organs
Ovaries	Estrogen and progesterone	Promote and maintain secondary sex characteristics

Pituitary Gland

The **pituitary gland** is composed of two parts, the anterior and posterior pituitary. It lies at the base of the brain and is connected to the **hypothalamus,** a portion of the brain that has centers for body temperature, sleep, sexual activity, and emotional states. The hypothalamus controls both the anterior and posterior pituitary. In fact, it produces the two hormones, **antidiurectic hormone (ADH),** also called vasopressin, and **oxytocin,** which are secreted by the posterior pituitary. These hormones pass from the hypothalamus to the posterior pituitary by way of nerve fibers (fig. 6.4). Oxytocin is of interest because it is involved in lactation. It causes milk letdown when and if the breast is producing milk, and it also stimulates uterine contraction (chapter 9). Researchers also believe that oxytocin may be involved in causing orgasm, as explained in the reading, "The Love Hormone."

The hypothalamus produces **releasing hormones** that control the anterior pituitary. These hormones, which pass from the hypothalamus to the anterior pituitary by way of tiny blood vessels, that is, capillaries, stimulte the anterior pituitary to produce and secrete its hormones. Each of the five major hormones made by the anterior pituitary has at least an alternative name composed of two parts: the first part indicates a target organ or gland, and the last part is the word *tropic,* which means "being drawn to" (table 6.1). Since the anterior pituitary produces hormones that regulate other endocrine glands, it is sometimes called the master gland (table 6.2, fig. 6.5).

Adrenal Glands

The **adrenal glands,** located atop the kidneys, have an outer cortex and inner medulla. Under the influence of **adrenocorticotropic hormone (ACTH)** the cortex secretes cortisol, a very important hormone involved in cellular metabolism. But of more interest to us is the fact that the

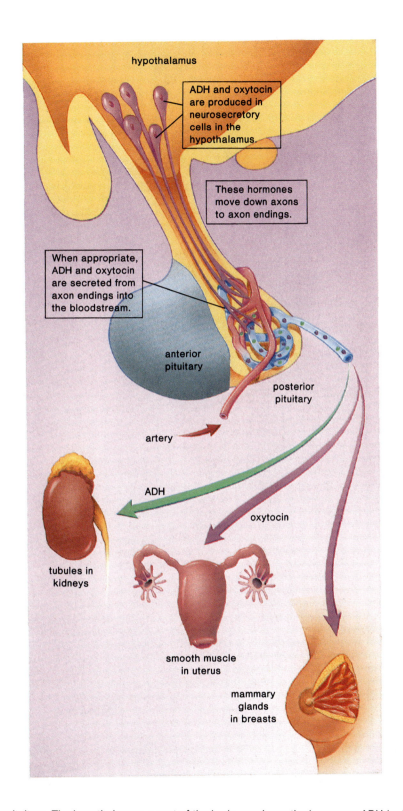

Figure 6.4

The actions of the posterior pituitary. The hypothalamus, a part of the brain, produces the hormones ADH (antidiuretic hormone) and oxytocin in neurosecretory cells that have processes (axons) leading to the posterior pituitary. The posterior pituitary releases these hormones, when appropriate into the bloodstream. ADH causes the kidneys to retain fluid; oxytocin causes the uterus (womb) to contract and the mammary glands in the breasts to release milk after the birth of a child.

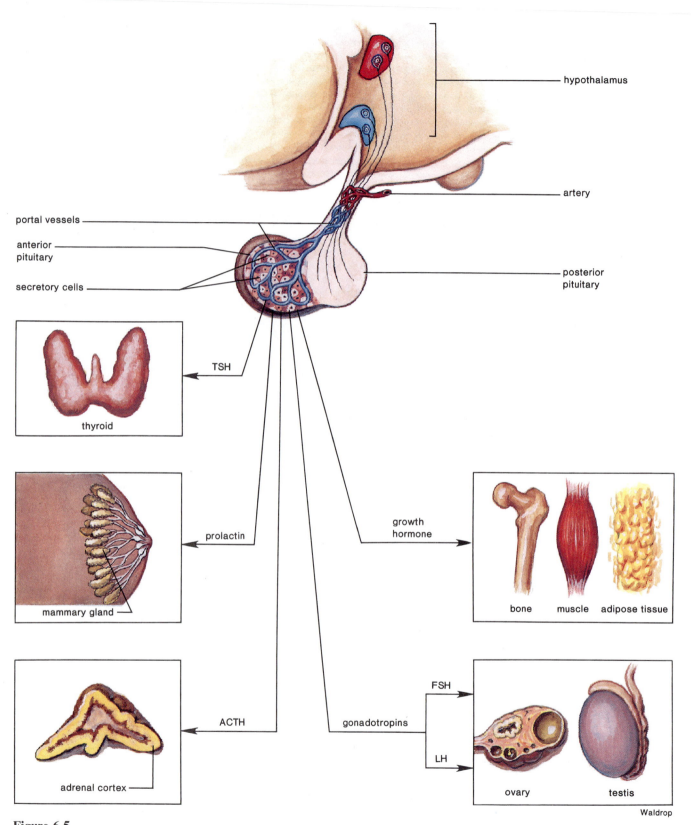

hypothalamus

artery

portal vessels

anterior
pituitary

secretory cells

posterior
pituitary

TSH

thyroid

prolactin

mammary gland

ACTH

adrenal cortex

growth
hormone

bone   muscle   adipose tissue

gonadotropins

FSH

LH

ovary   testis

Waldrop

Figure 6.5

The hypothalamus sends releasing hormones to the anterior pituitary by a circulatory route. The releasing hormones specifically promote or inhibit the secretion of anterior pituitary hormones.

The Love Hormone

Your irrationally romantic après-sex sentiments may be caused by oxytocin, a primary sexual arousal hormone that triggers orgasm and may make you feel emotionally attached to your partner. Prof. Niles Newton at Northwestern University Medical School was the first to hypothesize that oxytocin is "The Hormone of Love."

Oxytocin, a hormone that stimulates the smooth muscles and sensitizes the nerves, is produced by sexual arousal—the more you're in the mood, the more oxytocin you'll produce. As you begin to make love, oxytocin is released throughout your body, making your nerves more sensitive to pleasure and giving you that turned-on feeling. As the hormone level builds, oxytocin causes the nerves in your genitals to fire spontaneously, bringing on orgasm and giving you the feeling of losing control.

However, this substance does much more than simply activate your physical response to sex. Unlike other hormones, oxytocin arousal is generated by emotional cues, in addition to physical stimuli. The look he gives you, the sound of his voice or touch of his hand can all start the ball rolling, making oxytocin rush through your body (did you ever notice that you sometimes feel a wave of love come over you?), and may ultimately turn what started out as pure, unadulterated lust into a full-fledged feeling of love.

Naturally, when the object of your affections is Mr. Right, there's nothing wrong with this. The intriguing thing about oxytocin is that it can become conditioned to your own personal love history so that it's the sound of your particular partner's voice—not just anybody's—that sets you off.

But, while oxytocin works the same way in men, women seem to be more susceptible to its emotional effects. As you probably know all too well, women are more likely to feel a need for a continued connection after sex and are more vulnerable, as the cliché goes, to "loving too much." The reason for this: According to a Stanford University study led by Marie Carmichael, Ph.D., women have higher levels of oxytocin than men do during sex. Apparently, it takes more oxytocin for women to achieve orgasm (one reason it could take women longer to reach orgasm than men do). And with such a large amount of oxytocin in your body, it's no wonder you have an excess of emotion. Furthermore, when you break up with your boyfriend, your oxytocin level is likely to build up, sparking your need to be near him and making you instinctively want to get him back. Although this is painful, it can also feel exciting. And it's one explanation for why some women most want what they can't have—and keep returning to bad relationships. But keep in mind that there is one advantage: As a result of the oxytocin overload, women may be more capable than men are of having multiple orgasms and whole-body orgasms.

So don't worry if you feel too attached to Mr. Not-Quite-Right. It may not be love—it might be purely chemical.

From Charlotte Modahl, Ph.D., "The Love Hormone," November 1990. This article was originally published in *Mademoiselle*. Reprinted by permission of Dr. Charlotte Modahl.

TABLE 6.2	HORMONES PRODUCED BY THE ANTERIOR PITUITARY	
Hormone	**Gland stimulated**	**Hormone produced**
ACTH	Adrenal cortex	Cortisol
TSH	Thyroid gland	Thyroxin
FSH, LH	Gonads	
	Testes	Testosterone
	Ovaries	Estrogen and progesterone

adrenal cortex also produces a small amount of both male and female sex hormones. Therefore, in males the cortex is a source of female sex hormones, and in females it is a source of male hormones. It is possible that estrogen in males and androgen in females have important functions, but it is not known specifically what these functions are. If by chance the adrenal cortex begins to produce a large amount of sex hormones, it can lead to feminization in the male and musculinization in the female.

Gonads

The **gonads**—testes in the male and ovaries in the female—are controlled by two **gonadotropic hormones,** follicle-stimulating hormone (FSH) and luteinizing hormone (LH), which are produced by the anterior pituitary. These hormones are named for their action in the female, but they exist in both sexes, stimulating the appropriate gonads in each. In males, LH is sometimes called interstitial cell stimulating hormone (ICSH). The testes produce the male sex hormones, called **androgens,** of which the most potent is **testosterone.** The ovaries produce the female sex hormones, called **estrogen** and **progesterone.** The level of sex hormones in the body, or, for that matter, the level of most hormones, is regulated by a mechanism called feedback control.

Regulatory Control of Sex Hormones

By means of **feedback control** a system can regulate itself provided that it has a sensing device. Consider a home furnace and thermostat by which the temperature of individual rooms can be maintained using feedback control. The furnace produces heat, and when the temperature of a room reaches the temperature indicated on the thermostat, the sensing device in this case, the furnace is automatically shut off. On the other hand, when the temperature falls below that indicated on the thermostat, the furnace produces heat again (fig. 6.6). The production of heat by a furnace must be regulated, because it is not known in advance how the outside temperature will affect how much heat will be needed to maintain the desired temperature of the room.

Similarly, the need for hormones may fluctuate depending on the body's need. When the level of a hormone is sufficient, the gland producing the hormone is shut off; when the level of the hormone is insufficient, the gland begins to produce it again.

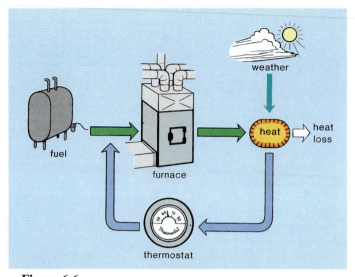

Figure 6.6

Feedback control. A furnace produces heat, and when the heat reaches a certain level, the thermostat (a sensing device) turns off the furnace. When heat is needed, the thermostat turns on the furnace. Since the product (heat) is controlling its own production, this is termed feedback control.

Sex hormone levels are maintained by a feedback control system that involves the hypothalamus, the anterior pituitary, and the gonads (fig. 6.7). The hypothalamic-releasing hormone called the **gonadotropic-releasing hormone (GnRH)** stimulates the anterior pituitary to produce the gonadotropic hormones, FSH and LH, which travel to the gonads by way of the circulatory system. In the male, the testes produce testosterone and in the female the ovaries produce estrogen and progesterone. These hormones promote the development of the primary and secondary sex characteristics, but they also exert feedback control over the hypothalamus and anterior pituitary. When the sex hormones reach a certain level in the body, the hypothalamus stops producing releasing hormones and the pituitary stops producing gonadotropic hormones. When the level of sex hormones falls, then, the hypothalamus again produces GnRH, reactivating the system.

As indicated in figure 6.7, the hypothalamus can also be controlled by the level of gonadotropic hormones in the body. In other words, both the level of sex hormones and the level of gonadotropic hormones serve to stimulate or depress the production of GnRH by the hypothalamus as appropriate.

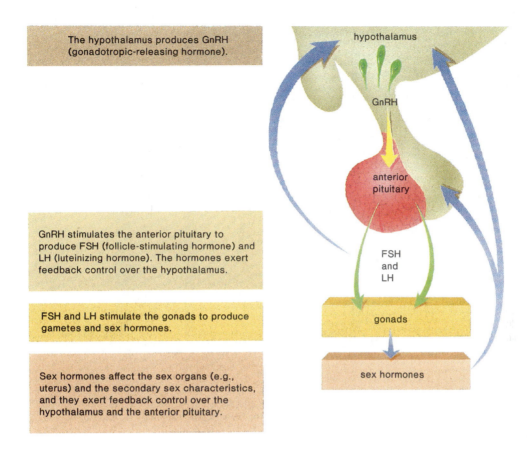

The hypothalamus produces GnRH (gonadotropic-releasing hormone).

hypothalamus

GnRH

anterior pituitary

GnRH stimulates the anterior pituitary to produce FSH (follicle-stimulating hormone) and LH (luteinizing hormone). The hormones exert feedback control over the hypothalamus.

FSH and LH

FSH and LH stimulate the gonads to produce gametes and sex hormones.

gonads

Sex hormones affect the sex organs (e.g., uterus) and the secondary sex characteristics, and they exert feedback control over the hypothalamus and the anterior pituitary.

sex hormones

Figure 6.7

The hypothalmic-pituitary-gonad system. This three-tiered system functions to control the levels of gonadotrophic and sex hormones in the blood.

SECONDARY SEX CHARACTERISTICS

At the time of puberty (ages 9–14), the sex organs mature and the secondary sex characteristics begin to appear. The cause of puberty is related to the level of sex hormones in the body. It is now recognized that the hypothalamic-pituitary-gonad system functions long before puberty, but the level of hormones is low because the hypothalamus is supersensitive to feedback control (fig. 6.8). At the start of puberty, the hypothalamus becomes less sensitive to feedback control and begins to increase its production of GnRH, causing the pituitary and the gonads to increase their production of hormones. The sensitivity of the hypothalamus continues to decrease until the gonadotropic and sex hormones reach the adult level.

The sex hormones, in conjunction with other hormones, have a profound effect on the body during puberty (fig. 6.9). There is an acceleration of growth leading to differences between the sexes in muscular and skeletal development, skin and hair growth, depth of voice, and breast development.

Muscular and Skeletal Sex Differences

Males commonly experience a growth spurt later than females, and they grow for a longer period of time (fig. 6.10). This means that males are generally taller than females and have broader shoulders and longer legs relative to

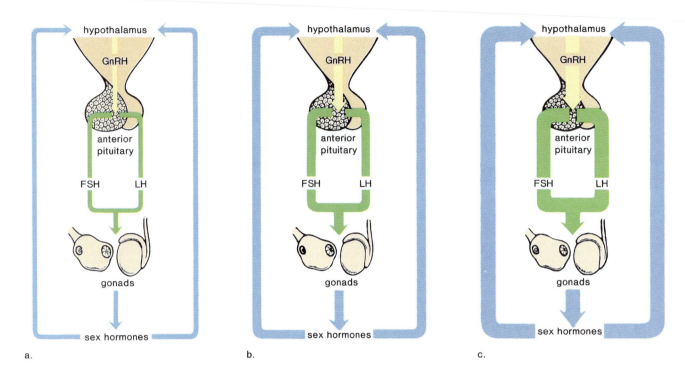

Figure 6.8

a. Before puberty the level of sex hormones is low because the hypothalamus is very sensitive to feedback control (blue arrows). *b.* At puberty the sensitivity of the hypothalamus decreases, and therefore the level of sex hormones increases in the body. Decreased sensitivity continues until (*c*) the adult level of sex hormones is reached.

trunk length. Androgens are responsible for the greater muscular development in males. Knowing this, males and females sometimes take anabolic steroids (either the natural or synthetic form of testosterone) to build up their muscles. The reading "Side Effects of Steroids" discusses the drawbacks to taking steroids for this purpose. On the other hand, estrogens are responsible for a greater accumulation of fat beneath the skin in females. This causes females to have a more rounded appearance than males.

The **pelvis** is a bony cavity formed by the hipbones and the sacrum and coccyx at the back (fig. 6.11). In females, a pelvic cavity outlet that can accommodate the passage of a baby's head allows a natural birth. During puberty, the pelvic girdle enlarges in females so that the pelvic cavity usually has a larger relative size compared to males. This means that females have wider hips than males and that the thighs converge at a greater angle toward the knees. Because the female pelvis tilts forward, females tend to have protruding buttocks, a more pronounced lower back curve than men, and an abdominal bulge.

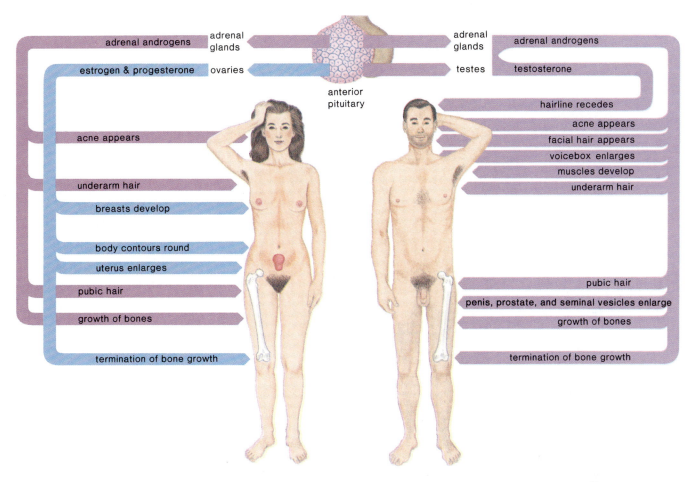

adrenal androgens — adrenal glands
estrogen & progesterone — ovaries

adrenal glands — adrenal androgens
testes — testosterone

anterior pituitary

hairline recedes
acne appears — acne appears
facial hair appears
voicebox enlarges
muscles develop
underarm hair — underarm hair
breasts develop
body contours round
uterus enlarges
pubic hair — pubic hair
penis, prostate, and seminal vesicles enlarge
growth of bones — growth of bones
termination of bone growth — termination of bone growth

Figure 6.9

Secondary sex characteristics of females and males. In the female, the anterior pituitary controls the secretion of adrenal androgens by the adrenal glands, and estrogen and progesterone, the female sex hormones, by the ovaries. The androgens have the effects indicated by the purple arrows and the female sex hormones have the effects noted by the blue arrows. In the male, the anterior pituitary controls the secretions of androgens by the adrenal cortex and androgens (i.e. testosterone) by the testes. These male sex hormones, together, have all the effects noted by purple arrows.

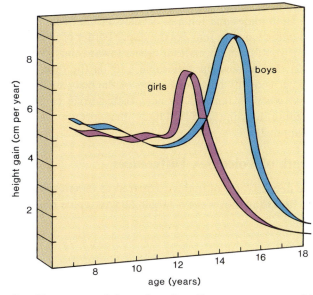

Figure 6.10

The growth spurt tends to occur earlier in females than in males. This means that girls are taller than males at age 12 or so. However, boys grow for a longer length of time and therefore they are taller in the end.

Sex Hormones and Secondary Sex Characteristics 81

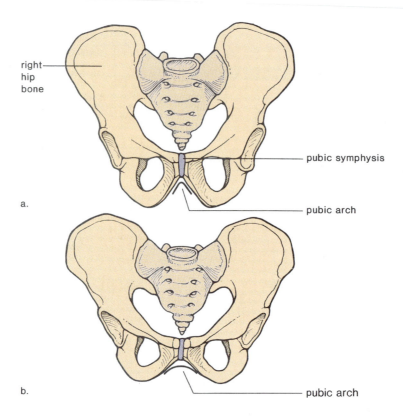

right hip bone

pubic symphysis

pubic arch

a.

b.

pubic arch

Figure 6.11

Human pelvic girdle. *a.* The pelvic girdle in males is not as wide as that of females and therefore males tend to have narrower hips. *b.* The pelvic girdle in females is wider than that of males and therefore females tend to have wider hips. This can be related to the need of a certain size opening to allow for the passage of a baby's head at birth.

Skin and Hair Sex Differences

The skin of females is softer and gentler than that of males. Oil- and sweat-producing glands become active in both sexes, and acne may accompany puberty due to clogged oil-producing glands.

Both males and females develop axillary (underarm) and pubic hair.[1] In females the upper border of pubic hair is horizontal, and in males it tapers toward the navel (fig. 6.12). Males develop noticeable hair on the face, chest, and occasionally on other regions of the body such as the back, whereas females do not. Testosterone causes the receding hairline and baldness that occurs in males (chapter 7).

Depth of Voice Sex Differences

The deeper voice of males compared to females is due to the fact that they have a larger larynx (voice box) with

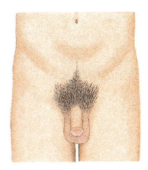

Figure 6.12

In females, the upper border of pubic hair is usually horizontal, while in males pubic hair tapers toward the navel.

longer vocal cords. Since the Adam's apple is a part of the voice box, it is usually more prominent in males than in females.

1. Pubic or pubis refers to the front of the pelvic girdle.

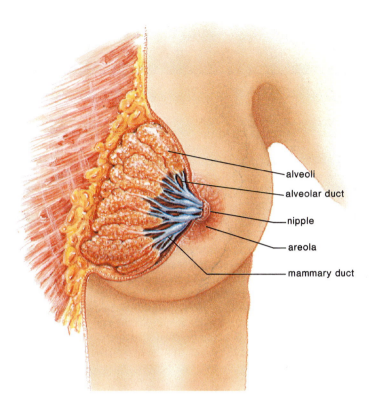

alveoli
alveolar duct
nipple
areola
mammary duct

Figure 6.13

The female breast contains lobules consisting of ducts and alveoli. The alveoli, lined by milk-producing cells, are more numerous in the lactating as opposed to nonlactating breast.

Breast Development in Females

Early growth of the breasts, or mammary glands, is referred to as "budding" of the breasts. Budding is followed by development of *lobes,* the functional portions of the breast, and deposition of adipose (fat) tissue, which gives breasts their adult shape.

A breast contains 15 to 25 lobes, each with its own milk duct that opens at the nipple. The nipple is surrounded by a pigmented area called the **areola.** Hair and sweat glands are absent from the nipples and areola, but glands are present that secrete a saliva-resisting lubricant to protect the nipples, particularly during nursing. Smooth muscle fibers in the region of the areola may cause the nipple to become erect in response to sexual stimulation or cold.

Within each lobe, the *mammary duct* divides into numerous *alveolar ducts* that can end in blind sacs called *alveoli* (fig. 6.13). The alveoli are made up of the cells that can produce milk.

Estrogen and progesterone are required for lobe development. It is believed that estrogen causes proliferation of ducts and that both estrogen and progesterone bring about alveolar development. The abundance of these hormones during pregnancy means that the alveoli proliferate at this time. There are ducts but few alveoli in the nonlactating breast, and there are many ducts and alveoli in the lactating breast. The role of prolactin (lactotropic hormone), an anterior pituitary hormone, in bringing about lactation and the role of oxytocin, a hormone secreted by the posterior pituitary, in causing milk letdown are discussed in chapter 9. ■

Being a steroid user may cost an athlete far more than his or her Olympic medal: a growing body of medical evidence indicates that athletes who take steroids have experienced problems ranging from sterility to loss of libido, and the drug has been implicated in the deaths of young athletes from liver cancer and a type of kidney tumor. Steroid use has also been linked to heart disease. "Athletes who take steroids are playing with dynamite," says Robert Goldman, 29, a former wrestler and weight lifter who is now a research fellow in sports medicine at Chicago Osteopathic Medical Center and who has published a book on steroid abuse, *Death in the Locker Room* (Icarus). "Any jock who uses these drugs is taking chances not just with his health but with his life."

Anabolic steroids are essentially the male hormone testosterone and its synthetic derivatives. . . .

The great majority of physicians say the drugs upset the body's natural hormonal balance, particularly that involving testosterone, which is present, though in different amounts, in both men and women. Normally, the hypothalamus, the part of the brain that regulates many of the body's functions, "tastes" the testosterone levels; if it finds them too low, it signals the pituitary gland to trigger increased production. When the hypothalamus finds the testosterone levels too high, as it does in the case of steroid abusers, it signals the pituitary to stop production. Problems can also arise in some cases after athletes stop taking the drugs and the hypothalamus fails to get the system started again.

The results can be traumatic. Many men experience atrophy, or shrinking of the testicles, falling sperm counts, temporary infertility and a lessening of sexual desire; some men grow breasts, while others may develop enlargement of the prostate gland, a painful condition not usually found in men under 50. Women who take too many steroids can develop male sexual characteristics. Some grow hair on their chests and faces and lose hair from their heads; many experience abnormal enlargement of the clitoris. Some cease to ovulate and menstruate, sometimes permanently.

There are several other health risks. Steroids can cause the body to retain fluid, which results in rising blood pressure. This often tempts users to fight "steroid bloat" by taking large doses of diuretics. A postmortem on a young California weight lifter who had a fatal heart attack after using steroids . . . showed that by taking diuretics he had purged himself of electrolytes, chemicals that help regulate the heart.

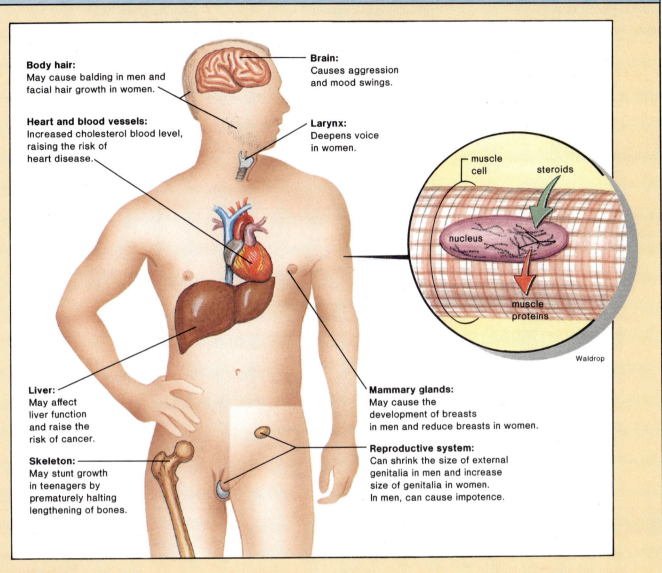

Body hair:
May cause balding in men and facial hair growth in women.

Heart and blood vessels:
Increased cholesterol blood level, raising the risk of heart disease.

Liver:
May affect liver function and raise the risk of cancer.

Skeleton:
May stunt growth in teenagers by prematurely halting lengthening of bones.

Brain:
Causes aggression and mood swings.

Larynx:
Deepens voice in women.

muscle cell

steroids

nucleus

muscle proteins

Waldrop

Mammary glands:
May cause the development of breasts in men and reduce breasts in women.

Reproductive system:
Can shrink the size of external genitalia in men and increase size of genitalia in women. In men, can cause impotence.

Figure 6.A
Steroids' effects on the body, including suspected harmful effects from anabolic steroid abuse; and how steroids build muscle (see insert).

SUMMARY

The secondary sex characteristics are controlled by hormones, notably the sex hormones produced by the gonads. Sex hormones, which act at the cellular level, are believed to turn on certain genes. Production is controlled by a feedback system involving the hypothalamus and anterior pituitary.

Increasing desensitivity at puberty causes the hypothalamus to increase its production of gonadotropic-releasing hormone (GnRH), which causes the anterior pituitary to increase its production of gonadotropic hormones.

The gonadotropic hormones stimulate the gonads to produce the sex hormones, with the subsequent development of the secondary sex characteristics involving distribution of hair, voice depth, breast development, and pelvic size.

KEY TERMS

adrenal glands 74

adrenocorticotropic hormone
(ACTH) 74

androgens 78

areola 83

antidiuretic hormone (ADH) 74

endocrine glands 71

estrogen 78

feedback control 78

gonadotropic hormones 78

gonadotropic-releasing hormone
(GnRH) 78

gonads 78

hormone 71

hypothalamus 74

oxytocin 74

pelvis 80

pituitary gland 74

primary sex characteristics 71

progesterone 78

puberty 71

releasing hormones 74

secondary sex characteristics 71

testosterone 78

REVIEW QUESTIONS

1. What role do sex hormones play in the embryonic development of the sex organs and the appearance of the secondary sex characteristics at puberty?

2. Explain how sex hormones work at the cellular level.

3. Define a hormone and name the endocrine glands involved in sex hormone production and regulation.

4. Describe the anatomical relationship of the pituitary gland to the hypothalamus and how the hypothalamus "controls" the pituitary gland.

5. Name the gonadotropic hormones produced by the anterior pituitary, and explain why the anterior pituitary may be called the master gland.

6. Describe the principle of feedback control.

7. How does feedback control operate in relation to the hypothalamic-pituitary-gonad system?

8. What are the secondary sex characteristics in males and females? When do they develop? What factors cause them to develop?

9. Describe the anatomy of the breast. Name the hormones needed for breast development and milk production and letdown.

CRITICAL THINKING QUESTIONS

1. Muscles contain two proteins, actin and myosin. Using the information provided in chapters 4 and 6, outline the precise steps by which you expect testosterone to increase muscle size.

2. An XX individual has inherited a genetic defect so that cellular receptors for estrogen are lacking.

   a. Would you expect this individual to have testes or ovaries? Why?

   b. Can this individual respond to estrogen? Why or why not? What affect will this have on the body?

   c. Taking into account that the adrenal cortex produces some testosterone, might this individual have the secondary sex characteristics of a male?

   d. Would you expect this individual to be fertile? Why or why not?

Male Reproductive Anatomy

*I*n humans, two gametes, the sperm and the egg, contribute chromosomes to the new individual. The sperm are small and swim to the egg, which is a much larger cell that contributes cytoplasm and organelles to the zygote. It seems reasonable that there should be a large number of sperm to ensure that a few will find an egg. In humans, the testes continually produce a plentiful supply of sperm. The testes (also called testicles) are a part of the male reproductive system, whose organs are listed in table 7.1 and depicted in figure 7.1.

TESTES

The **testes** are paired, oval-shaped organs measuring about 2 inches in length and 1 inch in diameter. The testes have two functions: they produce sperm, the male sex cells, and androgens, the male sex hormones.

The testes begin their development inside the abdominal cavity, but under the influence of testosterone they descend into the scrotal sacs, within the **scrotum,** which lies outside the body. If by chance the testes do not descend, a male has **cryptorchidism.** If he is not treated by administration of testosterone or operated on to place the testes in the scrotum, sterility follows. This type of

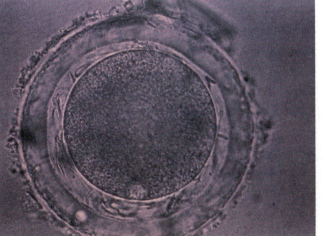

Human embryo undergoing first cell division.

sterility, the inability to produce offspring, occurs because normal sperm production does not occur at body temperature; a cooler temperature is required. Assuming that a male is wearing loose clothing, the scrotal sacs will hold the testes close to the groin or away from the groin as appropriate to maintain a temperature of about 95° F. The sacs can do this because they have layers of muscle that automatically adjust their degree of contraction in response to surrounding temperatures without need of conscious control. Tight clothing that squeezes the scrotum against the body can deter normal sperm production.

Seminiferous Tubules

A testis is divided into lobules (fig. 7.2). Each lobule contains one to three tightly coiled **seminiferous tubules** separated by interstitial cells. Altogether, these tubules have a combined length of approximately 750 feet. A microscopic cross section through a tubule shows that it is packed with cells undergoing spermatogenesis (fig. 7.2*b* and *c*). These cells are derived from undifferentiated germ cells, called *spermatogonia,* which lie just inside the testis and undergo mitosis (see fig. 1.4), always producing new spermatogonia.

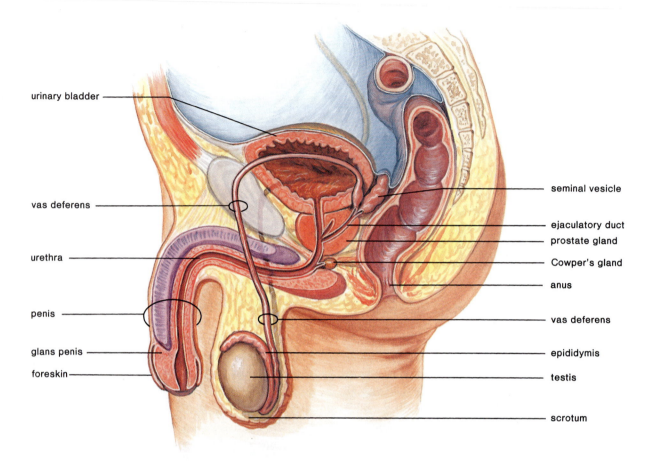

urinary bladder

vas deferens

urethra

penis

glans penis

foreskin

seminal vesicle

ejaculatory duct

prostate gland

Cowper's gland

anus

vas deferens

epididymis

testis

scrotum

Figure 7.1

Side view of the male reproductive system. See table 7.1 below.

TABLE 7.1	MALE REPRODUCTIVE SYSTEM
Organ	**Function**
Testis	Produces sperm and sex hormones
Epididymis	Maturation and some storage of sperm
Vas deferens	Conducts and stores sperm
Seminal vesicle	Contributes to seminal fluid
Prostate gland	Contributes to seminal fluid
Urethra	Conducts sperm
Cowper's gland	Contributes to seminal fluid
Penis	Organ of copulation

HUMAN ISSUE

Most public schools now have some sort of sex education program, and these have wide acceptance because of our concerns over child abuse, teenage pregnancy, AIDS, and other sexually transmitted diseases. Sex education, however, raises a number of controversial issues, such as in which grade it should begin and how explicit the course should be. For example, should the course include a description of birth-control methods and devices, and should all types of sex acts be discussed? If so, should the discussion confine itself to the medical consequences of certain acts or should it also include opinions on the morality of these behaviors? Where do you stand on these issues?

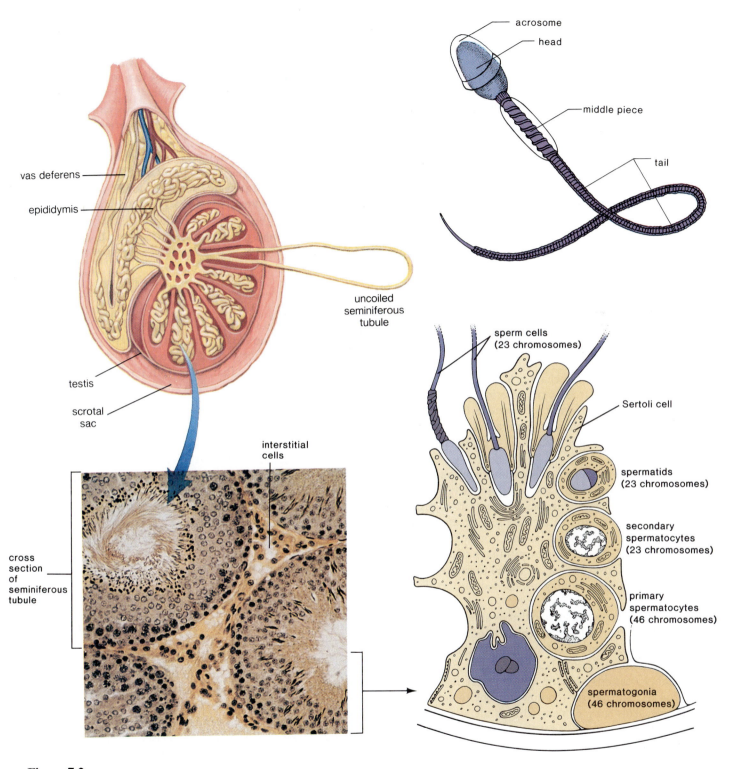

Figure 7.2

Anatomy of testis and sperm. *a.* A testis lies within a scrotal sac and contains many long and coiled seminiferous tubules (one uncoiled tubule is shown). The testis is connected to the epididymis which leads to the vas deferens. *b.* Light micrograph of cross section of tubules separated by cells called interstitial cells. *c.* Diagrammatic representation of spermatogenesis which occurs in tubules. (Compare to figure 1.10.) Note the Sertoli (nurse) cell which supports and protects the spermatocytes, spermatids, and finally the sperm. *d.* Mature sperm have a head, a middle piece, and a tail. The nucleus is in the head, capped by the enzyme-containing acrosome.

Germ cell is a generalized term for the cells in the testes and ovaries that give rise to the sex cells. Some spermatogonia increase in size, becoming *primary spermatocytes* that undergo meiosis, the type of cell division described in figure 1.8. As illustrated in figure 1.10, primary spermatocytes have 46 duplicated chromosomes, which pair and then separate to give *secondary spermatocytes,* each with 23 duplicated chromosomes. Secondary spermatocytes divide to give *spermatids,* which also have 23 chromosomes, but these are singled chromosomes. Spermatids then differentiate into spermatozoa, or mature sperm cells.

Notice in figure 7.2c that developing germ cells lie progressively farther away from the outer wall as they mature. Spermatogonia are near the wall, followed by spermatocytes and spermatids. Sperm are in the center of the tubule with their tails projecting into the internal cavity. Throughout the entire process of spermatogenesis, the germ cells are in intimate contact with **Sertoli,** or nurse, **cells** that apparently assist and nourish the germ cells as they mature. All together spermatogenesis takes 16 days to complete.

As we discussed in chapter 6, spermatogenesis is believed to be stimulated by the anterior pituitary hormone known as **follicle-stimulating hormone** (FSH), which is named for its function in females rather than in males. In a healthy male, sperm production is continuous from puberty to death.

The mature **sperm,** or spermatozoan, has three distinct parts: a head, a middle piece, and a tail (fig. 7.2d). The tail contains contractile filaments that propel the sperm at a rate of up to 1 inch per second, and the middle piece contains energy-producing organelles. The head contains the 23 chromosomes within a nucleus. The tip of the nucleus is covered by a cap called the **acrosome,** which is believed to contain enzymes needed for fertilization to take place.

The human egg is surrounded by several layers of cells and a membrane called the zona pellucida. The acrosome enzymes are believed to forge a path so that the sperm can reach the surface of the egg and enter. Each acrosome probably contains so little enzyme that it requires the action of many sperm to allow just one to actually enter the egg. This may explain why so many sperm are required for the process of fertilization. The normal human male may produce several hundred million sperm per day, assuring an adequate number for fertilization to take place (fig. 7.3).

Interstitial Cells

The male sex hormones, the androgens (e.g., testosterone), are produced by **interstitial cells,** which lie between the seminiferous tubules (fig. 7.2b). Testosterone promotes the maturation of the sex organs (table 7.1) and the development and maintenance of the secondary sex characteristics in males (chapter 6). Testosterone production is under the direct control of the anterior pituitary gonadotropic hormone LH (luteinizing hormone). Since LH is named for its action in females, it is sometimes given the name **interstitial cell stimulating hormone (ICSH)** in the male.

In general, a younger man produces more testosterone than an older man (fig. 7.4). Men may begin to exhibit decreasing sexual functions in their late forties and fifties. This decrease is called the male climacteric and may at times be accompanied by hot flashes, feelings of suffocation, and psychic disorders similar to those experienced by women during the menopause (chapter 8).

SPERMATIC DUCTS

The spermatic ducts include the epididymides (singular, epididymis), the vasa deferentia (singular, vas deferens), and the ejaculatory ducts. These ducts conduct sperm to the urethra, a tube that lies within the penis.

Epididymis

Sperm are made in the testes, but they undergo final maturation in the **epididymis,** which is a tightly coiled tubule about 20 feet in length that lies just outside the testes (fig. 7.1). Maturation seems to be required in order for the sperm to be capable of swimming and thus able to fertilize the egg. Also, it is possible that defective sperm are processed and destroyed during the 2 to 4 days that the sperm normally reside in the epididymides.

Vas Deferens

Each epididymis joins with a **vas deferens,** a tube about 18 inches long that leaves the scrotum and ascends through a canal called the *inguinal canal* into the abdominal cavity. In the abdominal cavity the vas deferens curves around the bladder to open into an ejaculatory duct. Sperm are stored for a matter of days in the first part of the vas deferens, which when stimulated undergoes rhythmical contraction to move the sperm into an ejaculatory duct.

The testes are suspended in the scrotum by the *spermatic cords,* each of which consists of connective tissue and muscle fibers enclosing the vas deferens, blood vessels, and nerves. The region of the inguinal canal, where the spermatic cord passes into the abdomen, remains a weak point in the abdominal wall. As such, it is frequently the site of a hernia, which is an opening or separation of some part of the abdominal wall through which a portion of internal organs, usually the intestine, is apt to protrude (fig. 7.5).

Figure 7.3
Scanning electron micrograph showing the single egg surrounded by the numerous and smaller sperm.

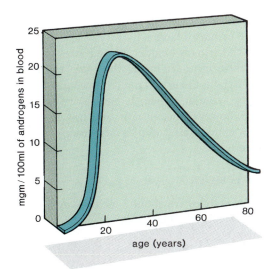

Figure 7.4
Sexual disfunction in older males may be due to decreasing level of testosterone. Younger men produce more testosterone than older men. Some men begin to exhibit decreasing sexual functions in their late forties and fifties because of a decreasingly lower level of testosterone.

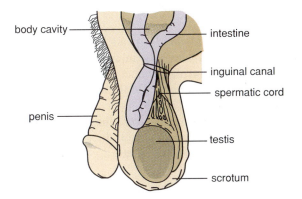

Figure 7.5
Anatomical defect in males due to an inguinal hernia. An inguinal hernia occurs when a loop of the intestine protrudes into a scrotal sac. This can be corrected surgically.

91

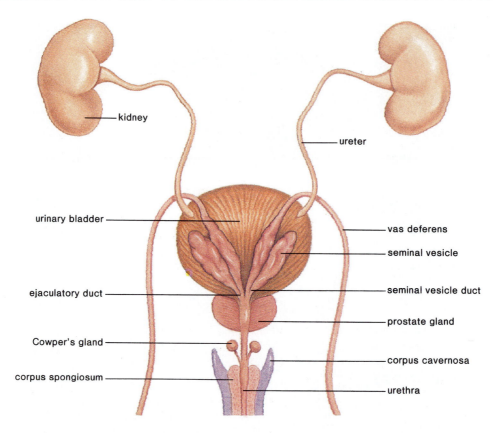

Figure 7.6

Accessory glands contribute secretions to semen discharged from the penis at ejaculation. A *seminal vesicle* (and vas deferens) on each side empty into an ejaculatory duct. The *prostate gland* surrounds and secretes into the urethra. The *Cowper's glands* secrete a substance that lines the urethra just before ejaculation occurs. Notice that in males the term urogenital system is apt because there is a connection between the urinary and reproductive systems.

Ejaculatory Duct

An **ejaculatory duct,** which is about 1 inch long, is formed after a vas deferens joins with a duct from a seminal vesicle (fig. 7.6). The ejaculatory duct actually passes through the prostate gland to empty into the urethra, whose major portion is in the penis. The ejaculatory ducts receive sperm from the vas deferentia and secretions from the seminal vesicles and prostate gland. Each duct ejects the sperm and these secretions into the urethra.

ACCESSORY GLANDS

Figure 7.6 shows the location of the accessory organs: the seminal vesicles, the prostate gland, and Cowper's glands. The secretions of these organs mixed with sperm has a milky appearance and is called **seminal fluid (semen).** Semen is conveyed by the urethra, through the penis, to the outside.

In the urinary system, urine, made by the kidneys, goes to the urinary bladder by way of the ureters. After storage in the urinary bladder, urine passes to the outside by way of the urethra. In males, therefore, the urethra has a function in both the urinary and reproductive systems— it can carry urine or semen. It never carries both at the same time, however, because at the time of ejaculation, the bladder exit is closed off by a circulatory muscle called a sphincter.

Seminal Vesicles

The **seminal vesicles** are about 2 inches long and lie at the base of the bladder. They empty their contents into the ejaculatory duct shortly after the vas deferens empties the sperm. The seminal vesicles secrete a sticky, slightly basic, yellowish substance that primarily (1) adds 60% of the bulk to semen, (2) provides a fluid medium for sperm, and (3) provides nutrients for sperm. Among other molecules,

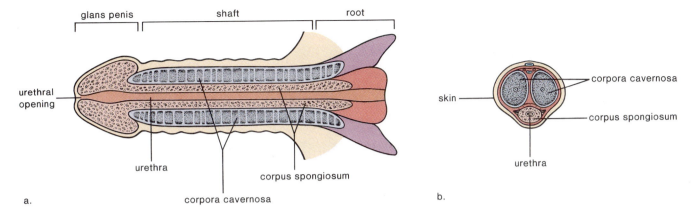

Figure 7.7

Anatomy of penis, the copulatory organ of males. *a*. The penis has three parts: the glans penis, the shaft, and the root. There are three columns of erectile tissue in the penis: two are the corpora cavernosa and one is the corpus spongiosum, which extends into the glans penis. *b*. Cross section of the penis.

the secretion from the seminal vesicles contains amino acids and a sugar, which is an energy source for sperm.

Seminal vesicle secretion also contains prostaglandins, chemicals that cause the uterus to contract. Some scientists now believe that uterine contraction is necessary to propel the sperm, and that the sperm only swim when they are in the vicinity of the egg.

Prostate Gland

The **prostate gland** is a single, doughnut-shaped gland about the size of a chestnut that surrounds the upper portion of the urethra just below the bladder. In older men, the prostate may enlarge and cut off the urethra, making urination painful and difficult. This condition may be treated with medicines or surgically.

The prostatic secretion is thin and milky in appearance. Its primary functions are to add some bulk to semen and to assist sperm motility and fertility because of its basic nature. Sperm do not become optimally motile unless they are in a basic fluid as opposed to an acidic fluid. The female vagina tends to be acidic, and the secretion of the prostate is needed to change this environment to one that is more favorable to sperm longevity and motility.

The prostate also secretes an enzyme called acid phosphatase, which is a coagulating agent. This enzyme is often measured clinically to assess prostate function. It clots the semen possibly to help prevent leakage of sperm from the vagina. Also present is a decoagulator that unclots the semen again.

Cowper's Glands

Cowper's glands are pea-sized organs that lie posterior to the prostate on either side of the urethra (fig. 7.6). Prior to ejaculation, these glands secrete a mucoid substance that lines the urethra, neutralizing the urine residue in the urethra, and lubricates the tip of the penis. Often it is possible to observe the secretion of Cowper's glands at the tip of the penis before ejaculation occurs.

PENIS

The penis has three portions: the *root* of the penis lies inside the body beneath the pubic bone; outside the body the penis consists of a *tubular shaft* and an enlarged tip called the **glans penis** (fig. 7.7). At birth the penis is covered by a layer of skin called the **foreskin,** or prepuce. Gradually over a period of 5 to 10 years, the foreskin becomes separated from the penis and may be retracted. During this time there is a natural shedding of cells between the foreskin and penis. These cells, along with an oil secretion that begins at puberty, is called **smegma.** In the child no special cleansing method is needed to wash away smegma, but in the adult the foreskin can be retracted to do so.

Circumcision is the surgical removal of the foreskin soon after birth. The pros and cons of circumcision are discussed in the reading entitled "Circumcision."

The penis contains three longitudinal masses of erectile tissue (fig. 7.7). Two of these, the corpora cavernosa, are located above the urethra and extend from the glans

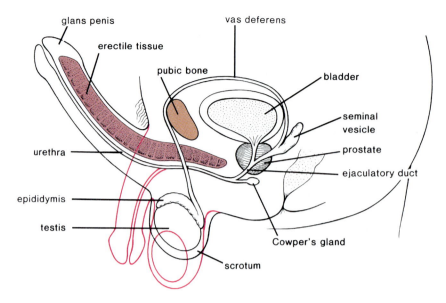

Figure 7.8

Erection occurs in males when the penis becomes capable of sexual intercourse. The position of the erect as opposed to the flaccid (nonerect) penis. The drawing shows only one column of erectile tissue; actually there are three columns (see fig. 7.7).

penis to its root. The other mass of erectile tissue, the corpus spongiosum, surrounds the urethra and enters the glans. The penis is flaccid and relaxed when the erectile tissue is not filled with blood, but when the tissue is filled with blood the penis becomes erect and capable of being placed into the vagina of a female (fig. 7.8).

Erection

Erection of the penis usually occurs when a male is sexually aroused (table 7.2 and fig. 7.9). During an erection, the penis becomes wider, longer, and firmer than the flaccid penis. Erection depends on the amount of blood entering the erectile tissue of the penis. When the penis is flaccid, the amount of blood entering the erectile tissue by way of small blood vessels called arterioles is the same as the amount of blood leaving by way of venules. But when the penis becomes erect, more blood enters than leaves the erectile tissue, and the erectile tissue becomes congested with blood. This is an example of *vasocongestion*, which, in general, is considered to be an important sexual response.

TABLE 7.2	**PHASES IN MALE SEXUAL RESPONSE**

Excitement Phase
Erection of the penis resulting from increased blood flow (vasocongestion). Penis increases in length and diameter. Partial elevation of testes and increase in size.

Plateau Phase
Increase in circumference of glans penis. Full elevation of testes. Appearance at tip of penis of mucoid material from bulbourethral glands (Cowper's glands).

Orgasmic Phase (Orgasm)
Ejaculation of semen resulting from contractions of the vas deferens and accessory organs. Contraction of anal and urethral sphincters.

Resolution Phase
Reduction in vasocongestion. Loss of penile erection. Refractory period sets in (i.e., inability to have another orgasm). May last for a few minutes or hours.

From Edwin B. Steen and James H. Price, *Human Sex and Sexuality,* 2d edition. Copyright © 1988 Dover Publications, Inc., New York, NY. Reprinted by permission.

The discussion regarding the advisability or inadvisability of circumcising the penis has centered on the following considerations (fig. 7.A).

Religion. In Islam and in Judaism, circumcision is a religious practice that represents a covenant with God made by Abraham.

Cleanliness. In preteen males, a number of small glands located in the foreskin and under the corona of the glans begin to produce an oily secretion. This secretion, along with dead skin cells, forms a cheesy substance known as smegma. In the circumcised male, smegma does not build up; in the uncircumcised, mature male, it collects, and its removal requires routine hygienic care. It is only necessary to retract the foreskin and wash away the smegma. If smegma is allowed to collect, bacteria can multiply and it will develop a strong odor. Also, the bacteria can cause a variety of infections in the female vaginal tract.

Function. Whether or not the foreskin protects the glans penis or whether it increases or decreases sexual sensitivity has not been definitely determined. The foreskin retracts during sexual intercourse, and no study has shown that its presence or absence has any affect on male performance or sensitivity.

Penile and Cervical Cancer. There is evidence that uncircumcised males may be slightly more at risk for developing penile cancer. For example, penile cancer almost never occurs among Jewish males.

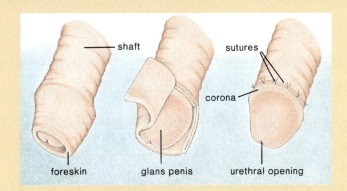

Figure 7.A
Circumcision is the removal of the foreskin so that the glans penis is exposed.

There is no evidence that the presence of a foreskin contributes to the development of cervical cancer in females. A study of Lebanese Muslims and Christians showed no greater incidence among the Christians compared to the Muslims, who practice circumcision.

Urinary Infections. It is not necessary to clean beneath the foreskin in newborn males; indeed the foreskin does not retract in newborn males and it should not be forced back. But several studies have reported a higher incidence of urinary infections among male infants. In one study it was found that 1,661 out of 40,000 infants developed a urinary infection. Among these infants, uncircumcised males had twice the incidence of urinary tract infections compared to girls and ten times the rate of circumcised boys. It is believed that in infants who wear diapers, bacteria do find a haven beneath the foreskin and thereafter can enter the urethra.

Surgical Trauma. Duration of crying and increase in heart rate suggests that circumcision is painful to newborns. Indications of pain are reduced when a local anesthetic is given. In rare instances hemorrhage, infection, mutilation, and even death have occurred. Whether or not there is long-lasting psychological trauma has not been definitely established. Sometimes circumcision is necessary later on in life due to phimosis, a condition in which the foreskin tightens and cannot be pulled back.

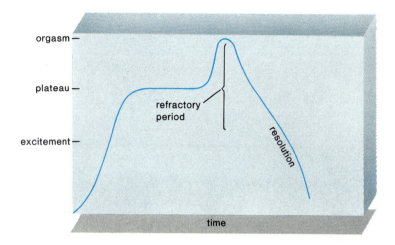

Figure 7.9

Male sexual response phases depicted graphically. For the events of each phase see table 7.2. Although ejaculation occurs during orgasm, orgasm is the physiological and psychological changes that occur as a result of release from muscular tension.

Erection is controlled by two portions of the central nervous system; sexual response centers are located in the spinal cord and in the brain. When a man thinks sexual thoughts or has sexual dreams, nerve impulses travel from the brain to a center in the spinal cord, which in turn emits nerve impulses that cause dilation of the arterioles leading to the erectile tissue. Conscious thought is not required for an erection, and stimulation of the penis alone can cause an erection. In this case, erection occurs due to a simple reflex action in which touch receptors in the penis initiate nerve impulses that are received by the spinal cord, and thereafter, nerve impulses leaving the spinal cord cause the arterioles to dilate and carry more blood. (Since an erection can occur automatically, it sometimes occurs unexpectedly.)

The scrotum also responds to sexual stimuli. Contraction of the muscles in the wall of the scrotum elevates the testes, which may have increased slightly in size.

The size of the flaccid penis is actually unimportant, because a small penis enlarges to a greater extent than a large penis. Also, the muscles at the base of the vagina constrict to adjust its size to that of the penis. In any case, a female receives stimulation through friction of the penis against the walls of the exterior opening of the vagina.

If the penis fails to become erect, the condition is called **impotency.** Although it was formerly believed that almost all cases of impotency were due to psychological reasons, it is now known that there are various medical reasons for impotency, such as hormonal imbalance, diabetes, and poor arterial circulation. Various types of medical remedies are available.

Ejaculation

If sexual arousal reaches its peak, **ejaculation** follows an erection. The first phase of ejaculation is called *emission.* During emission, the spinal cord sends nerve impulses via appropriate nerve fibers to the epididymides and vasa deferentia. Their subsequent motility causes sperm to enter the ejaculatory duct, whereupon the seminal vesicles, prostate gland, and Cowper's glands release their secretions. At this time, a small amount of secretion from the Cowper's glands may leak from the end of the penis. Since this is a mucoid secretion, it is believed that this leakage may aid the process of intercourse by providing a certain amount of lubrication. During the second phase of ejaculation, called *expulsion,* rhythmical contractions of muscles at the base of the penis and within the urethral wall expel semen in spurts from the opening of the urethra. These rhythmical contractions are an example of release from *myotonia,* or muscle tenseness. Myotonia is another important sexual response.

An erection lasts for only a limited amount of time. The penis, which was formerly tumescent (erect), now undergoes detumescence (nonerect) and returns to its normal flaccid state. Following ejaculation, a male may typically experience a period of time, called the **refractory period,** during which stimulation does not bring about an erection.

The contractions that expel semen from the penis are a part of male **orgasm,** the physiological and psychological sensations that occur at the climax of sexual stimulation. The psychological sensation of pleasure is centered in the brain, but the physiological reactions involve the genital (reproductive) organs and associated muscles as well as the entire body. Marked muscle tension is followed by contraction and relaxation.

MALE SEXUAL RESPONSE

Investigation into human sexual responses has intensified during the past several years. Data are accumulating that may eventually enable a complete description of this response. In the meantime, the work of William H. Masters and Virginia E. Johnson has been reported widely. For the sake of describing sexual response in detail, these investigators divided sexual response into four phases: (1) *excitement phase,* (2) *plateau phase,* (3) *orgasmic phase,* and (4) *resolution phase.* The events that could possibly be associated with each of these phases are given in table 7.2. It is important to realize that sexual response is ongoing and has arbitrarily been divided into these four phases. Also, the phase descriptions pertain to events that are believed to occur generally, and individual differences are to be expected. The same may be said for the male sexual response diagram proposed by Masters and Johnson (fig. 7.9). ■

SUMMARY

The gonads of a male are the testes, located in the scrotal sacs. Inside the testes, spermatogenesis occurs within seminiferous tubules, which are separated by interstitial cells, where the male sex hormones, for example, testosterone, are produced. Under the control of FSH produced by the anterior pituitary, spermatogenesis, which involves meiosis, occurs continuously. Consequently, the male produces a very

large number of sperm, each with 23 chromosomes. Testosterone production is under the control of ICSH, another anterior pituitary hormone.

Sperm mature in the epididymides and are stored primarily in the vas deferentia. At the time of sexual excitement, they move from vasa deferentia to the ejaculatory ducts that enter the urethra, located within the penis. Several glands (seminal vesicles,

prostate gland, and Cowper's glands) contribute fluid to semen, which passes from the penis during ejaculation. Four hundred million or more sperm may be found in semen.

The penis contains erectile tissue that fills with blood, enabling the penis to become erect so that sexual intercourse might take place.

KEY TERMS

acrosome 90
Cowper's glands 93
cryptorchidism 87
ejaculation 96
ejaculatory duct 92
epididymis 90
follicle-stimulating hormone (FSH) 90
foreskin 93
germ cell 90

glans penis 93
impotency 96
interstitial cell stimulating hormone (ICSH) 90
interstitial cells 90
orgasm 97
prostate gland 93
refractory period 97
scrotum 87

seminal fluid (semen) 92
seminal vesicles 92
seminiferous tubules 87
Sertoli cells 90
smegma 93
sperm 90
testes 87
vas deferens 90

REVIEW QUESTIONS

1. Name two functions of the testes and the part of the organ associated with each function.

2. Name the glands that contribute to semen. Are these glands endocrine glands?

3. State the path of sperm prior to and during ejaculation.

4. Give the normal sperm count per ejaculation. Offer reasons why the sperm count is so high.

5. What organ in males serves to transport both urine and semen?

6. Describe the structure of the penis.

7. Describe the processes of erection and ejaculation.

8. What are the four phases of sexual response? Relate the events of erection and ejaculation to these phases.

CRITICAL THINKING QUESTIONS

1. How could you redesign the male reproductive system so that the path of urine and the path of sperm are separate?

2. Animals that reproduce in water typically do not have a penis. What advantage does a penis serve for humans who reproduce on land? The penis has what disadvantage?

Female Reproductive Anatomy

*I*n humans, the female typically produces only one egg a month, and this one egg awaits the swimming sperm. Should fertilization occur, the developing embryo implants itself in the uterus, where development proceeds until birth occurs.

The sex organs of the reproductive system in females are listed in table 8.1 and depicted in figure 8.1.

OVARIES

In females, the **ovaries** are the sex organs that produce the female sex cells, called **eggs**, and the female sex hormones, estrogen and progesterone.

There are two ovaries, one on each side of the upper pelvic cavity. Each ovary ranges in size from 1 to 1½ inches in length and is anchored to the uterus and pelvic wall by ligaments. A longitudinal section through an ovary shows that it is made up of an outer *cortex*

Human embryo at 4-cell stage.

and inner *medulla*. There are many **follicles** in the cortex, and each follicle contains an oocyte (egg). A female is born with a large number of follicles (400,000 in both ovaries). Only a small number of these (about 400) will ever mature,

because a female produces only one egg per month during her reproductive years. Since oocytes are present at birth, they age as the woman ages. This has been cited as a reason why an older woman is more likely to produce a child with a genetic defect.

Oogenesis

As a follicle undergoes maturation, it develops from a primary follicle to a secondary follicle to a Graafian follicle. In a *primary follicle*, the *primary oocyte* is centrally placed, and it is enveloped by a single layer of follicular cells. The oocyte divides meiotically into two cells, each with 23 duplicated chromosomes (fig. 1.10). One of these cells, called the *secondary oocyte,* receives almost all the cytoplasm, nutrients, and enzymes. The other cell is a polar body that disintegrates or divides to give two polar bodies that die. There are many layers of follicular cells in the much larger *secondary follicle;* the secondary oocyte is pushed to one side of a fluid-filled cavity (fig. 8.2).

TABLE 8.1	FEMALE REPRODUCTION SYSTEM
Organ	**Function**
Ovary	Produces egg and sex hormones
Oviduct (fallopian tube)	Conducts egg
Uterus (womb)	Location of developing baby
Vagina	Copulatory organ and birth canal
External genitals	Sexual response

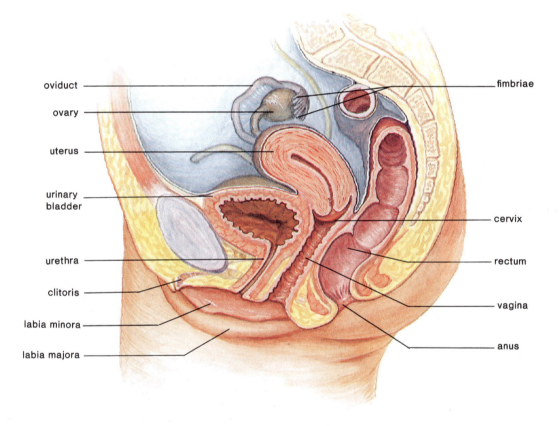

Figure 8.1
Side view of female reproductive organs.

A follicle that has reached maximum size is a **Graafian follicle.** In a Graafian follicle, the fluid-filled cavity increases to the point that the follicle wall balloons out on the surface of the ovary and bursts, releasing the secondary oocyte (often called an egg for convenience) surrounded by a thick membrane, the zona pellucida, and a few follicular cells, which together are called the corona radiata (fig. 1.11). This is referred to as **ovulation.** Com-pletion of oogenesis actually does not take place unless fertilization occurs. Fertilization, completion of oogenesis, and zygote formation occur in an oviduct.

Prior to ovulation, the follicle produces estrogen (and some progesterone). Following ovulation, the follicle is converted into the yellowish **corpus luteum,** which produces progesterone (and some estrogen). If fertilization does not take place, the corpus luteum degenerates into

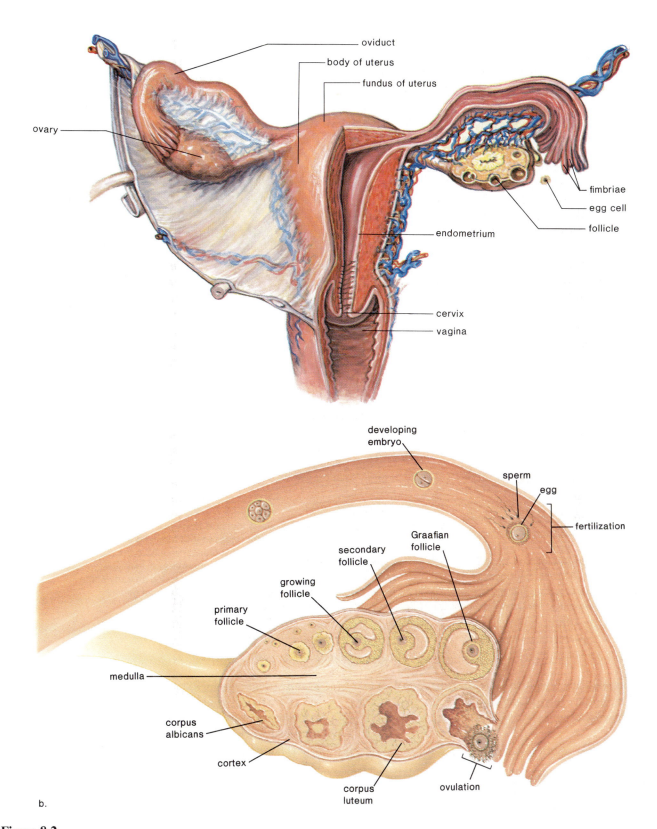

oviduct

body of uterus

fundus of uterus

ovary

fimbriae

egg cell

follicle

endometrium

cervix

vagina

developing embryo

sperm

egg

fertilization

Graafian follicle

secondary follicle

growing follicle

primary follicle

medulla

corpus albicans

cortex

corpus luteum

ovulation

b.

Figure 8.2

Female reproductive system. *a.* Front view. The ovaries are located on either side of the uterus adjacent to the fimbrae, fringed fingerlike processes, that sweep over the ovaries and direct the egg into an oviduct. *b.* Diagram of ovary showing the development of a follicle and egg. A follicle progresses from a primary to secondary to a Graafian follicle. The egg bursts from the follicle (and ovary) at ovulation, and then the follicle becomes the corpus luteum and finally the corpus albicans.

*T*wo embryos derived from one fertilized egg are identical twins (fig. 8.A). Fraternal twins, which are derived from two separate fertilized eggs, occur because of a multiple ovulation—more than one egg was released that month. Three factors that increase the chance of multiple ovulation are race (table 8.A), age, and the experience of previous pregnancies. An older woman who has already had several pregnancies is more likely to have a multiple ovulation than a young woman who has had no previous pregnancies.

Not only most twins but also triplets, quadruplets, quintuplets, and other multiple births are due to multiple ovulations. One cause of multiple ovulations is fertility drugs. Sometimes, infertile women are given hormones similar to esrogen to stimulate ovulation. One such drug, Clomid, scientifically known as clomiphene citrate, induces ovulation in roughly 70% of women who later conceive. If Clomid fails to bring on ovulation, doctors can prescribe Pergonal, which works in about 90% of women. Pergonal is HMG (human menopausal gonadotrophin) extracted from the urine of menopausal women. The anterior pituitary of menopausal women still produces the gonadotrophic hormones FSH and LH. Due to aging, the ovaries are unable to respond to the gonadotrophic

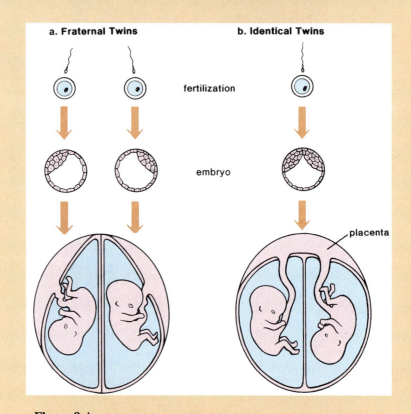

Figure 8.A

Conception of fraternal versus identical twins. *a.* Fraternal twins are formed when two eggs are released and fertilized. Fraternal twins receive a different genetic inheritance from both the mother and father. They can even have different fathers. *b.* Identical twins occur when the embryo breaks in two during an early stage of development. Identical twins have the exact same genetic inheritance from both the mother and father.

---

the **corpus albicans**, a whitish scar. If fertilization does take place, the corpus luteum persists for at most 6 months before degenerating.

Ovulation

Usually ovulation is singular, but multiple ovulations also occur as discussed in the reading entitled "Multiple Ovulations."

Sometimes women would like to know the day of ovulation in order to increase their chances of becoming pregnant or to prevent pregnancy from occurring. Ovulation usually occurs on the fourteenth day prior to the first day of the next menstrual bleeding. This is not terribly helpful information, because it is generally not known ahead of time just when the first day of the next bleeding will be.

It has been discovered that at ovulation, there is a clear and runny discharge that has the consistency of egg white. Before and after ovulation, the discharge is whitish and cloudy. Also, the body temperature changes at ovulation. Basal body temperature, which must be taken upon waking in the morning, falls slightly before ovulation and then rises immediately after ovulation (fig. 8.3).

TABLE 8.A   TWINNING IN DIFFERENT RACES

Race	Averate Rate per 1,000 Births	
	Monozygotic*	Dizygotic**
Negroids		
Africa	4.9	22.3
United States	3.9	10.9
Mongoloids		
Japan	4.1	2.3
Caucasoids		
Europe	3.6	8.6
United States	3.8	6.1

*Identical twins
**Fraternal twins
Table from *Human Biology* by John Cunningham. Copyright © 1983 by John Cunningham. Reprinted by permission of HarperCollins Publishers.

hormones, and therefore the ovarian and uterine cycles do not occur. When estrogen and progesterone are not produced by the ovaries, FSH and LH production is not inhibited, and instead these hormones are produced in excess of the usual amounts. The excess is excreted in the urine and can be extracted—this is the source of HMG.

The administration of HMG to a young woman may cause her to have multiple ovulations and consequently to experience multiple childbirths following a single pregnancy. In other words, a woman who was previously infertile becomes capable of giving birth to four or five children at a time because the fertility drug causes her to produce this number of eggs in one month.

HMG can also be given to couples who decide to use in vitro fertilization as a means of having a child (chapter 9). In that case several preovulatory eggs are surgically collected and subsequently fertilized in laboratory glassware before they are implanted in the uterus.

Since these methods for determining the time of ovulation are rather imprecise, they are usually more helpful when a woman is trying to become pregnant than when she is trying to prevent pregnancy. Still, these methods are discussed further in the section on birth control (chapter 10).

Ovarian Cycle

The monthly changes a follicle goes through is called the **ovarian cycle.** The ovarian cycle is under the control of the anterior pituitary gonadotropic hormones, **follicle-stimulating hormone (FSH)** and **luteinizing hormone (LH).** It is clear from the names of these hormones that FSH may be associated with the maturation of a follicle and that LH may be associated with the functioning of the corpus luteum.

The gonadotropic hormones are not present in constant amounts in the female, rather, they are secreted at different amounts during a monthly ovarian cycle that usually lasts 28 days but may vary widely (fig. 8.4). For simplicity's sake, it is most convenient to assume that during the *first half* of a 28-day cycle, the *follicular phase,* the anterior pituitary is secreting FSH, and the follicle in

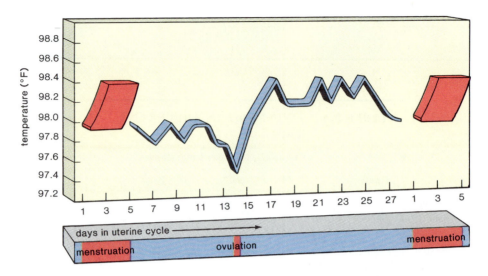

Figure 8.3

Typical oral body temperature changes during a 28-day uterine (menstrual) cycle. Temperature falls slightly before ovulation and then rises immediately after ovulation.

the ovary is secreting estrogen. Just as the hypothalamus has caused the anterior pituitary to begin producing FSH, so FSH has brought about follicle development in the ovary. As the blood estrogen level rises, it exerts feedback control over the hypothalamus and anterior pituitary so that this follicular phase comes to an end (fig. 8.5). A sudden surge of LH production causes ovulation to occur on the fourteenth day of a 28-day cycle.[1] During the *last half* of the ovarian cycle, the *luteal phase,* the anterior pituitary is producing LH, and the corpus luteum is secreting progesterone. The hypothalamus has caused the anterior pituitary to produce LH, which in turn promotes conversion of the follicle into the corpus luteum. As the blood progesterone level rises, it exerts feedback control over the hypothalamus and the anterior pituitary so that this luteal phase comes to an end. Then, the corpus luteum degenerates.

The female sex hormones, estrogen and progesterone, have numerous functions (table 8.2). Some of these functions were discussed in chapter 6. Later in this chapter, we will discuss the effect that these hormones have on the uterus.

OVIDUCTS

The **oviducts,** also called the uterine (or fallopian) tubes, are about 4 inches long and extend from the uterus to the ovaries (fig. 8.2). These muscular tubes are lined by cells

possessing cilia, which are little hairlike structures that beat toward the uterus. The tubes are not attached to the ovaries and instead have fingerlike projections called fimbria that sweep over the ovary at the time of ovulation. When the egg bursts from the ovary during ovulation, it is usually sucked up into an oviduct by the combined action of these projections and the beating of the cilia.

Since an egg must traverse a small space before entering an oviduct, it can possibly get lost and enter the abdominal cavity. Such eggs usually disintegrate, but rarely they are fertilized in the abdominal cavity and even more rarely have come to term, with the child delivered by surgery. Development of an embryo anywhere outside the uterus is called an **ectopic pregnancy.**

Once in the oviduct, the egg is at first propelled rapidly by cilia movement and tubular muscle contraction. Soon the contractions diminish, and the egg moves more slowly toward the uterus. As mentioned, fertilization, completion of oogenesis, and zygote formation occur in the oviduct. The developing embryo normally arrives at the uterus after several days and then embeds itself in the uterine lining, which has been prepared to receive it. Occasionally, an ectopic pregnancy occurs when the embryo becomes embedded into the wall of an oviduct, where it begins to develop. Tubular pregnancies cannot succeed because the tubes are not anatomically capable of allowing full development to occur.

1. Ovulation occurs on the fourteenth day prior to the first day of menstruation. Only in a 28-day cycle would this be the fourteenth day counting from day 1.

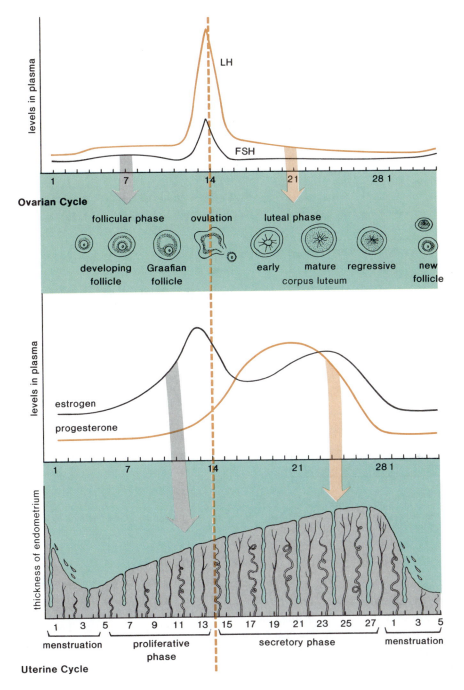

Figure 8.4

Ovarian cycle (above) and uterine cycle (below). Before ovulation, the anterior pituitary releases the gonadotropic hormone FSH, which stimulates the maturation of an ovarian follicle. A follicle primarily produces estrogen, which causes the endometrium to become thicker (proliferative phase of uterine cycle). After ovulation, the gonadotropic hormone LH stimulates maturation of corpus luteum. A corpus luteum primarily produces progesterone, which causes the endometrium to become secretory (secretory phase of uterine cycle). Due to feedback inhibition, a low level of all these hormones in the body brings on menstruation.

The hypothalamus produces GnRH (gonadotropic-releasing hormone).

GnRH stimulates the anterior pituitary to produce FSH (follicle-stimulating hormone) and LH (luteinizing hormone).

FSH stimulates the follicle to produce estrogen and LH stimulates the corpus luteum to produce progesterone.

Estrogen and progesterone affect the sex organs (e.g., uterus) and the secondary sex characteristics, and they exert feedback control over the hypothalamus and the anterior pituitary.

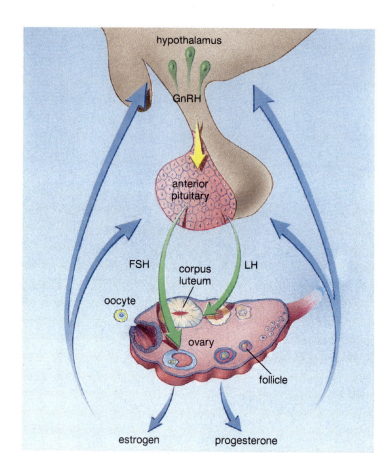

Figure 8.5

The hypothalamic-pituitary-gonad system in the female. The left half of the diagram particularly pertains to the first half of the month, and the right half particularly pertains to the second half of the month.

UTERUS

The single **uterus** is about the size and shape of an inverted pear and lies tipped forward over the urinary bladder between the bladder and the rectum (fig. 8.1). The organ is held in place by uterine ligaments, but even so its position changes according to the fullness of the urinary bladder and rectum (fig. 8.6). If the bladder is full, the uterus is moved backward; if the rectum is full, the uterus is moved more forward.

The muscular uterus has three portions: the fundus, the body, and the cervix (fig. 8.2). The oviducts join the uterus just below the fundus, and the cervix enters into the vagina at a nearly right angle. The opening of the cervix, called the os, leads to the vaginal canal.

Development of the embryo takes place in the uterus. This organ, sometimes called the womb, is capable of stretching to over 12 inches to accommodate the growing baby. The lining, called the endometrium, participates in

the formation of the placenta, which supplies the nutrients needed for embryonic and fetal development.

The endometrium has two layers: a basal layer and an inner functional layer. The functional layer of the endometrium varies in thickness according to a monthly cycle, called the **uterine cycle.** The uterine cycle is controlled principally by the ovarian hormones, estrogen and progesterone. With increased production of estrogen by an ovarian follicle, the endometrium thickens, becoming vascular and glandular. This is called the *proliferation phase*. With increased production of progesterone by the corpus luteum, the endometrium doubles in thickness and the uterine glands become mature, producing a thick, mucoid secretion. This is called the *secretory phase* of the uterine cycle. The endometrium is now prepared to receive the developing zygote, which usually becomes embedded in the lining several days following fertilization. During this process, called **implantation,** the deve-

TABLE 8.2 EFFECTS OF FEMALE SEX HORMONES

Estrogen

1. Growth of ovaries and follicles
2. Growth and maintenance of the smooth muscle and epithelial linings of the entire reproductive tract
   Also: a. Oviducts: increased motility and ciliary activity
         b. Uterus: increased motility secretion of abundant, clear cervical mucus
         c. Vagina: increased "cornification" (layering of epithelial cells)
3. Growth of external genitals
4. Growth of breasts (particularly ducts)
5. Development of female body configuration: narrow shoulders, broad hips, converging thighs, diverging arms
6. Stimulation of fluid sebaceous gland secretions ("anti-acne")
7. Pattern of pubic hair (actual growth of pubic and axillary hair is androgen-stimulated)
8. Sex drive and behavior (? role of androgens)
9. Reduction of blood cholesterol
10. Vascular effects (deficiency ⟶ "hot flashes")
11. Feedback effects on hypothalamus and anterior pituitary

Progesterone

1. Stimulation of secretion by endometrium; also induces thick, sticky cervical secretions
2. Stimulation of growth of myometrium (in pregnancy)
3. Decrease in motility of oviducts and uterus
4. Decrease in vaginal "cornification"
5. Stimulation of breast growth (particularly glandular tissue)
6. Inhibition of effects of prolactin on the breasts
7. Elevation of body temperature
8. Feedback effects on hypothalamus and anterior pituitary

From H. J. Vander, et al. *Human Physiology* p. 447. Copyright © 1975 McGraw-Hill, Inc., New York. Reproduced with permission of McGraw-Hill, Inc.

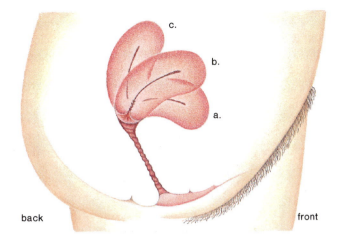

Figure 8.6

Approximate location of uterus in a standing woman with (*a*) bladder and rectum empty, (*b*) bladder and rectum full, and (*c*) full bladder and empty rectum.

TABLE 8.3 OVARIAN AND UTERINE CYCLES

Ovarian Cycle Phases	Events	Uterine Cycle Phases	Events
Follicular	FSH	Menstruation, days 1–5	Endometrium breaks down
Days 1–13	Follicle maturation		
	Estrogen	Proliferation, days 6–13	Endometrium rebuilds
OVULATION Day 14*			
Luteal	LH		
Days 15–28	Corpus luteum	Secretory Days 15–28	Endometrium thickens and glands secrete glycogen
	Progesterone		

*Assuming a 28-day cycle.

loping embryo becomes enclosed by the uterine lining, where development continues. The female is now pregnant.

If pregnancy does not occur, the corpus luteum begins to degenerate, progesterone level decreases, and the functional layer of the endometrium breaks down. Blood, cells, and glandular secretions flow from the uterus as the thickened layer is shed. This phase of the uterine cycle is called the **menstruation,** or the menses. Menstruation results in a discharge of 1 to 6 ounces. Before menstruation has completely ceased, repair begins under the influence of estrogen from the ovary, where follicular development is again underway.

The Uterine and Ovarian Cycles

The uterine and ovarian cycles are synchronized because both are ultimately under the control of the anterior pituitary hormones FSH and LH. Figure 8.4 and table 8.3

indicate how the ovarian cycle brings about the uterine cycle. Day 1 of both cycles is the first day of menstruation. Notice that the follicular phase of the ovarian cycle encompasses both the menstruation and proliferation phases of the uterine cycle. At this time, because of the presence of FSH, a follicle is developing in the ovary and is producing estrogen. This hormone brings about repair of the endometrium as menstruation is ceasing. Following ovulation, the luteal phase of the ovarian cycle corresponds to the secretory phase of the uterine cycle. At this time, because of the presence of LH, the corpus luteum is producing progesterone, and this hormone prepares the uterine lining to receive the zygote. If fertilization has not occurred or if the zygote is unable to implant itself, the corpus luteum degenerates and the uterine lining breaks down.

Some authorities believe that women experience mood changes according to the level of sex hormones and the corresponding time of the monthly cycle. Low estrogen levels in particular have been associated with such emotions as anxiety, tension, depression, and irritability, while high estrogen levels have been associated with confidence and cheerfulness. So many factors affect mood that studies in this regard have not been decisive.

Menarche and Menopause

The **menarche** is the first uterine cycle and **menopause** is the cessation of the uterine cycle. Menopause is likely to occur between the ages of 45 and 55. As menopause develops, the uterine cycles become irregular, but as long as they occur it is still possible for a woman to conceive and become pregnant. Therefore, a woman is usually not considered to have completed menopause until there has been no menstruation for a year. Menopause is associated with certain physical symptoms such as "hot flashes," which are caused by circulatory irregularities; dizziness; headaches; insomnia; sleepiness; and depression. Again, there is a great variation among women, and any of these symptoms may be absent altogether.

Women sometimes report an increased sex drive following menopause, and it has been suggested that this may be due to androgen production by the adrenal cortex or even by the ovaries themselves.

VAGINA

The **vagina** is a 5- to 6-inch muscular tube that makes a 45° angle with the small of the back (fig. 8.1). The lining of the vagina lies in folds called rugae, which are capable of extension as the muscular wall stretches. This capacity to extend is especially important when the vagina serves as the birth canal, and it may facilitate intercourse when the vagina receives the penis during copulation.

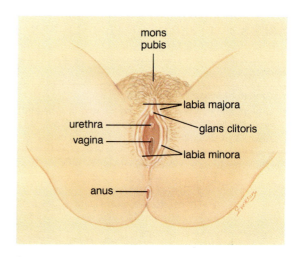

Figure 8.7
External genitals of a female.

Substances produced by cells in the lining of the vagina are metabolized to acids inside the vagina. These acids are believed to prevent bacterial infections of the genital tract. But they also make the vagina hostile to sperm, because sperm prefer a basic rather than an acidic environment. Since seminal fluid is basic it is capable of neutralizing the vagina.

EXTERNAL GENITALS

The external genital organs of the female are known collectively as the **vulva** (fig. 8.7). The **mons pubis** is a fatty prominence underlying the pubic hair. The **labia majora** are two large folds of skin that extend backward from the mons pubis. The outer sides of the labia majora are pigmented and have hair, but the inner sides are smooth and have numerous modified sweat glands. Since the labia majora develops from embryonic genital folds, as does the scrotum, they are considered homologous to the scrotum.

The **labia minora** are two small folds lying just inside the labia majora. They consist of smooth, pigmented skin and extend upward from the vaginal opening to encircle and form a foreskin for the **clitoris,** an organ that is homologous to the penis. Although quite small, the clitoris has a shaft of erectile tissue and is capped by a pea-shaped glans.

The **vestibule,** a cleft between the labia minora, contains the openings of the urethra and the vagina. The urethra is part of the urinary system, while the vagina is the copulatory organ of females. Notice then that the urinary system and the reproductive system are separate in the female. The vagina may be partially closed by a ring of tissue called the **hymen.** The hymen may persist or be disrupted by all types of physical activities, and not necessarily sexual intercourse; therefore, its presence or absence should not be associated with virginity.

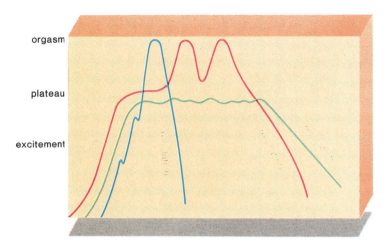

Figure 8.8

Female sexual response phases depicted graphically. For the events of each phase see table 8.4.

FEMALE SEXUAL RESPONSE

Sexual response in the female may be more subtle than in the male, but there are certain corollaries. Recall from the previous chapter that vasocongestion and release from myotonia (muscle tenseness) were important sexual responses.

The clitoris is believed to be an especially sensitive organ for initiating sexual sensations. It is possible for the clitoris to become slightly erect as its erectile tissues become engorged with blood. But vasocongestion is more obvious in the labia minora, which expand and deepen in color. Erectile tissue within the vaginal wall also expands with blood, and the added pressure in these blood vessels causes small droplets of fluid to squeeze through the vessel walls and lubricate the vagina. Another possible source of lubrication, especially in prolonged intercourse, is from the mucus-secreting Bartholin's glands located beneath the labia minora on either side of the vagina.

Not only is there swelling in the vagina and the labia minora, the uterus enlarges and elevates. Also, the breasts of women typically swell due to engorgement, and the nipples may become erect.

Release from myotonia occurs in females, especially in the region of the vulva and vagina but also throughout the entire body. Increased uterine motility may assist the transport of sperm toward the oviducts. Female *orgasm* is not signaled by ejaculation, and there is a wide range in normality regarding sexual response. Table 8.4 outlines and figure 8.8 diagrams the female sexual response according to four phases as proposed by Masters and Johnson. ∎

TABLE 8.4 PHASES IN FEMALE SEXUAL RESPONSE

Excitement Phase
Moistening of vagina by appearance of beads of moisture on inner surface. Increase in size of clitoris. Elevation of uterus.

Plateau Phase
Engorgement and swelling of tissues in outer third of vagina and labia. Ballooning of inner two-thirds of vagina. Further elevation of uterus and cervix. Enlargement of uterus; elevation of clitoris. Appearance of mucoid material from glands of Bartholin.

Orgasmic Phase (Orgasm)
Contraction of uterus, vagina, anal and urethral sphincters. (Nothing in the nature of an ejaculation occurs.)

Resolution Phase
Reduction in vasocongestion. Reduction in orgasmic platform. Decrease in size of clitoris. Lack of a refractory period with ready return to orgasm.

From Edwin B. Steen and James H. Price, *Human Sex and Sexuality,* 2d edition. Copyright © 1988 Dover Publishing, Inc., New York, NY. Reprinted by permission.

HUMAN ISSUE

How far should a couple be willing to go to have a good sexual relationship? If the relationship is not good should they simply ignore it? Or should they buy books and listen to radio and television programs about sex or even go to a sex therapist?

SUMMARY

The gonads of a female are the ovaries, which are located one on each side of the upper pelvic cavity. Inside the ovaries are numerous follicles. Each month one follicle matures completely, producing estrogen and an egg, which bursts from the ovary during ovulation. Then, the follicle becomes the corpus luteum, which produces the other female sex hormone, progesterone. Estrogen causes the uterine lining (endometrium) to build up, and progesterone causes the uterine lining to become secretory. The egg usually enters one of the two oviducts that lead to the uterus. The egg is fertilized within the oviduct before proceeding to the uterus, where the developing embryo implants itself in the endometrium lining.

If fertilization does not occur, the endometrium is shed during menstruation. The ovarian (and uterine cycle) has two parts: the follicular (proliferation) phase is under the control of the anterior pituitary hormone, FSH, which promotes maturation of a follicle, and the luteal (secretory) phase is under the control of LH, which maintains the corpus luteum. Feedback control can explain the occurrence of the two phases, including menstruation.

The vagina is the organ of copulation and the birth canal. Sexual response in females also involves the external genitals.

KEY TERMS

clitoris 108

corpus albicans 102

corpus luteum 100

ectopic pregnancy 104

eggs 99

follicle-stimulating hormone (FSH) 103

follicles 99

Graafian follicle 100

hymen 108

implantation 106

labia majora 108

labia minora 108

luteinizing hormone (LH) 103

menarche 108

menopause 108

menstruation 107

mons pubis 108

ovarian cycle 103

ovaries 99

oviducts 104

ovulation 100

uterine cycle 106

uterus 106

vagina 108

vestibule 108

vulva 108

REVIEW QUESTIONS

1. State the organs in the female reproductive system and give a function for each.

2. Describe the anatomy of the ovary and the oviducts.

3. Describe the events of the ovarian cycle and state the hormones associated with each event.

4. Where does oogenesis, ovulation, and fertilization occur? What is an ectopic pregnancy?

5. Describe the anatomy of the uterus and vagina.

6. Describe the events of the uterine cycle. How are the events of the uterine cycle related to the events of the ovarian cycle?

7. Name the external genitals of the female, and state their anatomical position in relation to each other.

8. How does the female sexual response compare to male sexual response?

CRITICAL THINKING QUESTIONS

1. How could you redesign the female reproductive system so that the vagina is not both an organ of copulation and the birth canal?

2. Following menopause, females stop producing eggs and cannot get pregnant. Males produce sperm their entire lives. Relate this discrepancy to differences in the reproductive function and expected behavior of the sexes.

Fertilization, Development, and Birth

Successful fertilization, development, and birth are required before a human being can come into existence.

FERTILIZATION

If intercourse is accompanied by ejaculation, some 200 to 600 million sperm are deposited at the rear of the vagina in the region of the cervix (fig. 9.1). The semigelatinous seminal fluid protects the sperm from the acid of the vagina for several minutes, after which they are killed unless they have managed to enter the uterus. Whether or not the sperm enter depends in part on the consistency of the cervical mucus. Three to four days prior to ovulation and on the day of ovulation, the mucus is watery, and the sperm can penetrate it easily. During the other days of the uterine cycle, the mucus is thicker and has a sticky consistency and the sperm can rarely penetrate it.

Fertilization normally occurs in the upper third of the oviduct, and a small percentage of sperm (perhaps 100,000) usually arrive there within 5 to 30 minutes. It is believed that uterine and oviduct contractions transport the sperm, and that the prostaglandins within seminal fluid may promote these contractions. Swimming action seems

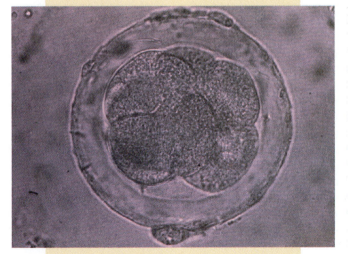

Human embryo at 8-cell stage.

to be important only when the sperm are in the vicinity of the egg.

Freshly ejaculated sperm are relatively inactive and unable to fertilize the egg. The sperm must undergo *capacitation* by being exposed to the female reproductive tract for several hours. Following capacitation, the sperm are more active. Also, when capacitated sperm contact the corona radiata surrounding an egg, acrosome enzymes are released that allow a sperm to penetrate the egg (fig. 9.2).

As soon as a single sperm head has made its way into the region of the cytoplasm, the zona pellucida undergoes a marked and rapid chemical transformation that makes it impossible for other sperm to enter. The sperm nucleus moves through the cytoplasm and fuses with the egg nucleus, which has just undergone the second meiotic division. Fertilization is now complete. If fertilization does not occur, the oocyte does not finish meiosis and disintegrates instead.

The fertilized egg, more properly called the **zygote,** has the diploid number of chromosomes and begins dividing. The developing embryo travels very slowly down the oviduct to the uterus, where it implants itself in the prepared uterine lining (fig. 9.2). Upon implantation, the female is pregnant.

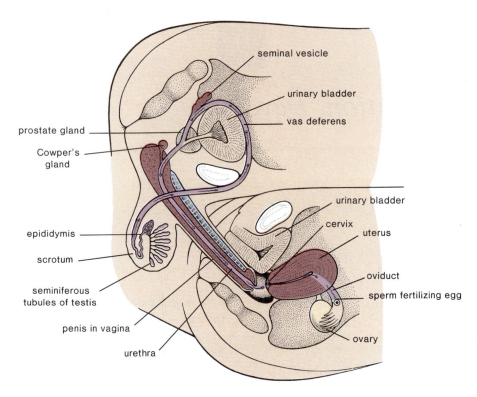

Figure 9.1

Longitudinal section of male and female pelvic regions showing the penis in the vagina at the time of ejaculation. In the male, the sperm travel along the vas deferens and ejaculatory duct before entering the urethra where they are joined by secretion from the accessory glands. In the female, the sperm pass through the cervix and the uterus before entering the oviduct, where fertilization occurs.

If pregnancy does occur, the uterine lining is maintained, and menstruation does not normally occur during the entire time of pregnancy. Menstruation ceases because the outer layer of cells surrounding the embryo, the chorion, which will be discussed following, produces the gonadotropic hormone HCG (human chorionic gonadotropin). HCG prevents the normal degeneration of the corpus luteum and instead stimulates it to secrete even larger quantities of progesterone. After its formation, the placenta, also discussed following, continues production of HCG and begins production of progesterone and estrogen. The latter hormones have two effects: they shut down the anterior pituitary so that no new follicles are started, and they maintain the lining of the uterus so that the corpus luteum is no longer needed.

The physical signs that might prompt a woman to have a pregnancy test are cessation of menstruation, increase in frequency of urination, morning sickness, and increase in the size and tenderness of the breasts, as well as the development of a dark coloration of the areolae.

Pregnancy tests are based on the fact that HCG is present in the blood and urine of a pregnant woman. Clinical urine tests are generally accurate by 10 to 14 days after a missed menstruation. Blood tests are more expensive but can indicate a pregnancy earlier, often before the time of a missed menstruation. Home pregnancy tests measure HCG in the urine and are generally quite reliable by 2 weeks after a missed menstruation.

DEVELOPMENT

Development encompasses the time from conception (fertilization followed by implantation) to birth (parturition). In humans the gestation period, or length of pregnancy, is approximately 9 months. It is customary to calculate the time of birth by adding 280 days to the start of the last menstruation because this date is usually known, whereas the day of fertilization is usually unknown. Because the time of birth is influenced by so many variables, only about 5% of babies actually arrive on the forecasted date.

Human development is very often divided into **embryonic development** (the first 2 months) and **fetal development** (the third through ninth months). The embryonic period consists of early development, during which all the major organs form, and fetal development consists of a

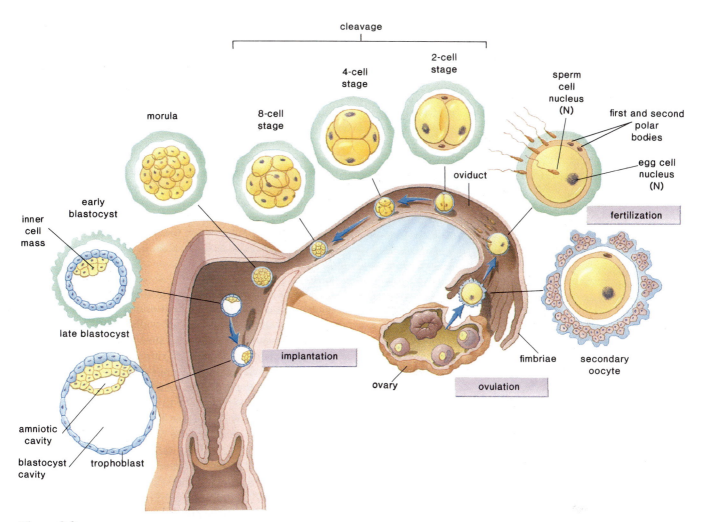

Figure 9.2

Human development before implantation. At ovulation, the secondary oocyte (egg) leaves the ovary. Fertilization occurs in the oviduct. As the zygote moves along the oviduct, it undergoes cleavage to produce a morula, a solid ball of cells. Rearrangement of cells gives a blastocyst having an inner cell mass (embryo) and trophoblast (outer layer of cells). The trophoblast becomes the extraembryonic membrane, the chorion. The amnion is also an extraembryonic membrane. The blastocyst implants itself in the uterine lining and the woman is now pregnant.

refinement of these structures. The fetus, not the embryo, is recognizable as a human being.

One of the major events in early development is the establishment of the extraembryonic membranes (fig. 9.3).

Extraembryonic Membranes

The term **extraembryonic membranes** is apt, because these membranes extend out beyond the embryo. One of the membranes, the **amnion,** provides a fluid environment for the developing embryo and fetus. It is a remarkable fact that all animals, even land-dwelling humans, develop in water. Amniocentesis, a process by which amniotic fluid, and the fetal cells floating therein, are withdrawn for examination, is described in the reading for this chapter. One

authority describes the functions of amniotic fluid in this way:

> The colorless amniotic fluid by which the fetus is surrounded serves many purposes. It prevents the walls of the uterus from cramping the fetus and allows it unhampered growth and movement. It encompasses the fetus with a fluid of constant temperature which is a marvelous insulator against cold and heat. Above all, it acts as an excellent shock absorber. A blow on the mother's abdomen merely jolts the fetus, and it floats away.[1]

1. A. F. Guttmacher, *Pregnancy, Birth and Family Planning* (New York: New American Library, 1974), p. 74.

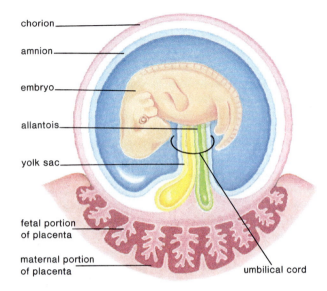

chorion

amnion

embryo

allantois

yolk sac

fetal portion
of placenta

maternal portion
of placenta

umbilical cord

Figure 9.3

Embryo surrounded by two extraembryonic membranes: the amnion and chorion. Two other extraembryonic membranes are the yolk sac and allantois.

The **yolk sac** is another extraembryonic membrane. Yolk is a nutrient material utilized by other animal embryos—the yellow of a chick's egg is yolk, for example. In humans, however, the fetus is not nourished by yolk, nor does it eat food. Instead, its nutritional needs are met by the extraembryonic membrane called the **chorion,** which is a part of the placenta (fig. 9.4). In humans, the yolk sac is the first site of red blood cell formation, and in that way the yolk sac contributes to the development of the circulatory system. Another extraembryonic membrane, the **allantois,** contributes even more; its blood vessels become umbilical blood vessels that take fetal blood to and from the placenta.

Placenta

The **placenta** has two portions: the fetal portion is chorionic tissue, and the maternal portion is uterine tissue (fig. 9.4). The placenta arises after the embryo has implanted itself in the uterine wall. Here the chorion is in intimate contact with maternal tissue, and the placenta comes into being.

Maternal blood never mingles with fetal blood at the placenta; instead, only nutrient and waste molecules are exchanged across blood vessel walls. The fetus receives oxygen and nutrient molecules such as glucose and amino acids. It gets rid of carbon dioxide and other wastes such as urea, our primary nitrogenous waste.

Note that the digestive system, lungs, and kidneys do not function in the fetus. The functions of these organs are not needed because the fetus has its nutritional and excretory needs supplied by the placenta.

Some **congenital** (birth) **defects** are genetic in origin. The presence of defective genes can be detected by using the methods discussed in the reading entitled "Detecting Birth Defects." Other congenital defects or illnesses are environmental in nature and are caused by substances that have crossed from mother to fetus at the placenta.

Medications and other drugs present in an expectant mother's blood do pass into fetal blood at the placenta. The most notorious example of the harmful developmental effects that can result from drugs are the thalidomide babies (fig. 9.5). An estimated 10,000 children were born with deformed arms and sometimes deformed legs because their mothers took this sedative during their pregnancies.

Drug users and those who overuse alcohol have babies who display withdrawal symptoms, and smokers have underweight babies. Because of these known effects, it seems reasonable to suggest that a pregnant woman should be very careful about taking drugs of any sort because of the

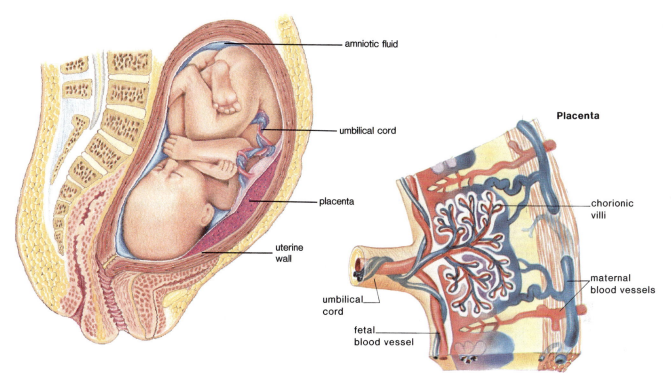

Figure 9.4

At the placenta, fetal circulation and maternal circulation lie in close proximity so that waste molecules pass from the fetal to the maternal circulation and nutrient molecules pass in the opposite direction.

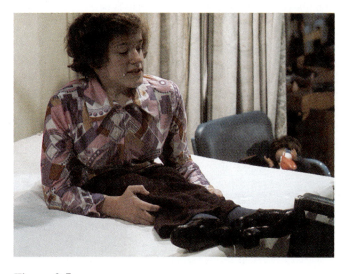

Figure 9.5

This girl's mother took thalidomide when she was pregnant. The drug crosses the placenta and causes deformed limbs.

effect they may have on her developing child. Also, pregnant women should avoid X-ray diagnosis. Dividing cells are particularly susceptible to damage by X ray, and, as we shall see, cell division is absolutely essential to development.

In addition to drugs, disease organisms can pass from the expectant mother's blood to her baby's blood. For example, an embryo can contract AIDS, toxoplasmosis, Lyme's disease, and syphilis from its mother in this way. Rubella (German measles) is also an early deformer of embryonic development. Particularly during the first 3 months of her pregnancy a German measles infection may cause congenital malformations, particularly of the heart, eyes, and ears.

Drugs and disease often affect embryonic rather than fetal development because this is the period of time when structures are first appearing. Unfortunately, this is also the time when women most likely do *not* realize that they are pregnant. Therefore, especially when birth control is not being practiced, the prospective mother should carefully watch her health and intake of drugs.

Umbilical Cord

While it may seem that the **umbilical cord** travels from the intestines to the placenta, actually the umbilical cord is simply taking fetal blood to and from the placenta (fig. 9.4). The umbilical cord is the lifeline of the fetus because it contains the umbilical arteries and vein, which transport waste molecules (carbon dioxide and urea) to the placenta for disposal and take oxygen and nutrient molecules from the placenta to the rest of the fetal circulatory system (fig. 9.6).

Fetal circulation has other features not seen in adult circulation. In the adult, for example, blood entering the right side of the heart passes to the lungs for oxygenation before returning to the left side of the heart. In the fetus, blood moves directly from the right to the left side of the heart by way of an opening called the oval opening (*foramen ovale*). Any blood that does happen to leave the right side of the heart still bypasses the lungs by entering a connector vessel called the arterial duct (*ductus arteriosus*).

Embryonic Development

Development, which begins with a single cell, proceeds from the simple to the complex. First, the single cell becomes a number of cells by the process of mitosis, then these cells differentiate into tissues, and finally the tissues become organized into organs. Development is an orderly process by which each preceding event seems to trigger the event that follows (table 9.1).

First Week

Immediately after fertilization, the zygote divides repeatedly as it passes down the oviduct to the uterus. The many cells arrange themselves so that there is an inner cell mass surrounded by a layer of cells, the trophoblast, which will become the chorion (fig. 9.2). The early appearance of the chorion emphasizes the complete dependence of the developing embryo on this membrane. The inner cell mass is the **embryo.**

Once it has arrived in the uterus on about the seventh day, the embryo begins to embed itself in the uterine lining (fig. 9.2).

Second Week

Implantation is completed during the second week. With implantation, pregnancy has now taken place, and the placenta begins formation. The ever-growing number of cells are now arranged in tissues; the amniotic cavity is seen above the embryo and the yolk sac is below (fig. 9.7a).

Third Week

Another extraembryonic membrane, the allantois, makes its appearance briefly, but later it and the yolk sac become part of the umbilical cord as it forms (fig. 9.7c and d). Organs are already developing, including the spinal cord and heart. The nervous system and circulatory system have begun.

TABLE 9.1 HUMAN DEVELOPMENT

Time	Events for Mother	Events for Baby
Embryonic Development		
First Week	Ovulation occurs.	Fertilization occurs. Cell division begins and continues. Chorion appears.
Second week	Symptoms of early pregnancy (nausea, breast swelling and tenderness, tiredness) are present.	Implantation. Amnion and yolk sac appear. Embryo has tissues. Placenta begins to form.
Third week	First missed menstruation. Blood pregnancy test is positive.	Nervous system begins. Allantois and blood vessels are present. Placenta is well formed.
Fourth week	Urine pregnancy test is positive.	Limb buds begin. Heart is noticeable and beating. Nervous system is prominent. Embryo has tail. Other systems begin.
Fifth week	Uterus is size of hen's egg. Frequent need to urinate due to pressure of growing uterus on bladder.	$\frac{1}{12}$ in. Embryo is curved. Head is large. Limb buds show divisions. Nose, eyes, and ears are noticeable.
Sixth week	Uterus is growing to size of an orange.	¼ in. Fingers and toes are present. Cartilaginous skeleton.
Two months	Uterus can be felt above the pubic bone.	½ in. All systems are developing. Bone is replacing cartilage. Refinement of facial features.
Fetal Development		
Third month	Uterus is the size of a grapefruit.	3 in (1 oz). Possible to distinguish sex. Fingernails.
Fourth month	Fetal movement is felt by those who have been pregnant before.	8.5 in (6 oz). Skeleton visible. Hair begins to appear.
Fifth month	Fetal movement is felt by those who have not been pregnant before. Uterus reaches up to level of umbilicus and pregnancy is obvious.	12 in (1 lb). Protective cheesy coating begins to be deposited. Heartbeat can be heard.
Sixth month	Doctor can tell where baby's head, back, and limbs are. Breasts have enlarged, nipples and areolae are darkly pigmented, and colostrum is produced.	14 in (2 lbs). Body is covered with fine hair. Skin is wrinkled and red.
Seventh month	Uterus reaches halfway between umbilicus and rib cage.	16 in (4 lbs). Testes descend into scrotum. Eyes are open.
Eighth month	Weight gain is averaging about a pound a week. Difficulty in standing and walking because center of gravity is thrown forward.	Body hair begins to disappear. Subcutaneous fat begins to be deposited.
Ninth month	Uterus is up to rib cage, causing shortness of breath and heartburn. Sleeping becomes difficult.	20 in (7 lbs). Ready for birth.

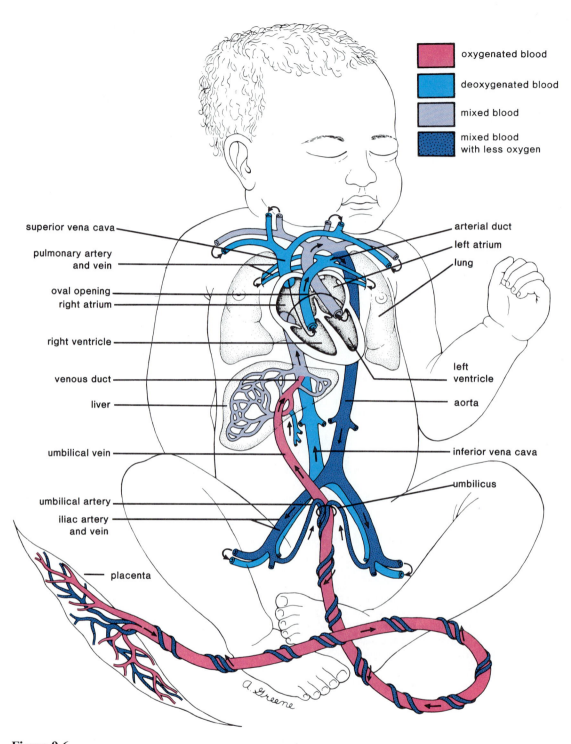

oxygenated blood

deoxygenated blood

mixed blood

mixed blood
with less oxygen

superior vena cava

pulmonary artery
and vein

oval opening
right atrium

right ventricle

venous duct

liver

umbilical vein

umbilical artery

iliac artery
and vein

placenta

arterial duct

left atrium

lung

left
ventricle

aorta

inferior vena cava

umbilicus

a. Greene

Figure 9.6

Fetal circulation. Blood is oxygenated at the placenta and enters the umbilical vein, which joins with the vena cava. From the vena cava, blood enters the right side of the heart and passes directly to the left side by way of the oval window, or else it enters the pulmonary trunk and passes to the aorta by way of the arterial duct. In either case, blood does not go to the lungs.

118

*T*hree methods for genetic defect testing before birth are amniocentesis, chorionic villi sampling, and screening eggs for genetic defects.

Amniocentesis cannot be done until the sixteenth week of pregnancy. A long needle is passed through the abdominal wall to withdraw a small amount of amniotic fluid along with fetal cells. Since there are only a few cells in the amniotic fluid, testing must be delayed for 4 weeks until cell culture produces enough cells for testing purposes.

Chorionic villi sampling can be done as early as the fifth week of pregnancy. The doctor inserts a long, thin tube through the vagina into the uterus. With the help of ultrasound, which gives a picture of the uterine contents, the tube is placed between the lining of the uterus and the chorion. Then suction is used to remove a sampling of the chorionic villi cells. Chromosome analysis and biochemical tests for several different genetic defects can be done immediately on these cells.

Screening eggs for genetic defects is the newest of these techniques. Preovulation eggs are removed by aspiration after a laparoscope (a telescope with a cold light source) is inserted into the abdominal cavity through a small incision in the region of the navel. The prior administration of FSH ensures that several eggs are available for screening. Only the chromosomes within the first polar body are tested, because if the woman is heterozygous for a genetic defect and it is found in the polar body, then the egg must be normal. Normal eggs undergo in vitro fertilization, discussed following, and are placed in the prepared uterus. At present, only one in ten attempts results in a birth, but it is known ahead of time that the child will be normal.

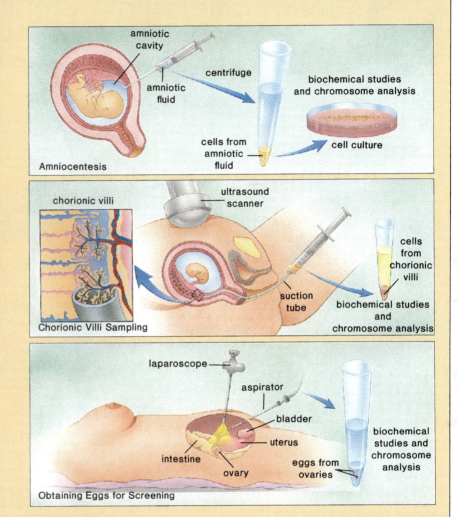

Figure 9.A
Methods for genetic defect testing before birth.

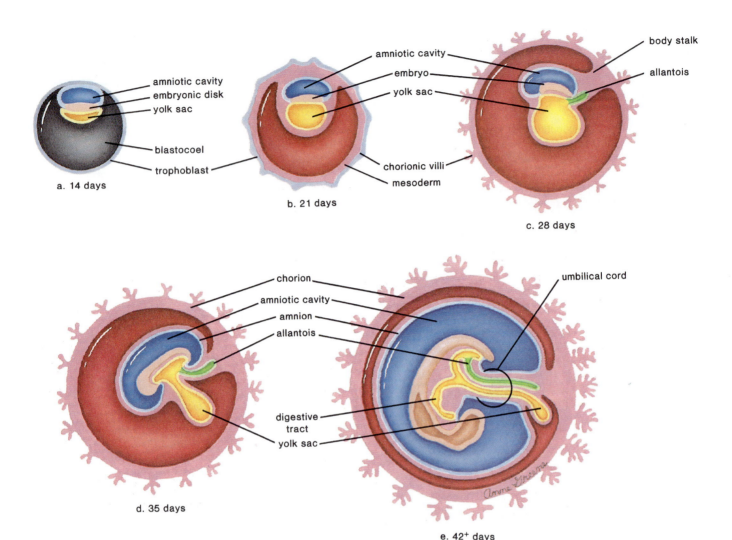

Figure 9.7

Embryonic development. *a.* At first there are only tissues present in embryo. The amniotic cavity is above the embryo, and the yolk sac is below. *b.* The chorion is developing villi so important to exchange between mother and child. *c.* The allantois and yolk sac are two more extraembryonic membranes. *d.* These extraembryonic membranes are positioned inside the body stalk as it becomes the umbilical cord. *e.* At 42 days the embryo has a head region and tail region. The umbilical cord takes blood vessels between the embryo and the chorion (placenta).

First Month

By the end of the first month, the placenta is producing enough human chorionic gonadotropic hormone (HCG) to maintain the corpus luteum and the uterine lining. The fetal portion of the placenta, the chorion, has treelike villi that project into the uterine tissue. As the fetus develops, these villi enlarge at one location only, although they may eventually take up as much as 50% of the uterus (fig. 9.4).

The embryo has a nonhuman appearance largely due to the presence of a tail, but also because the arms and legs, which begin as limb buds, resemble paddles. The head is much larger than the rest of the embryo, and the whole embryo bends under its weight (fig. 9.8). Eyes, ears, and

nose are just appearing. The enlarged heart beats and the bulging liver takes over the production of blood cells for blood that will carry nutrients to the developing organs and wastes from the developing organs.

Second Month

At the end of the second month, the embryo's tail has disappeared, and the arms and legs are better formed with fingers and toes apparent (fig. 9.9). The head is very large, the nose is flat, the eyes are far apart, and the ears are distinctively present. Internally, all major organs have appeared. Embryonic development is now finished.

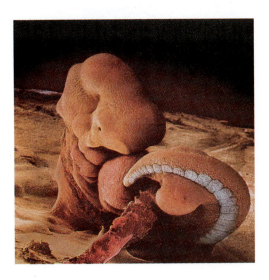

a.

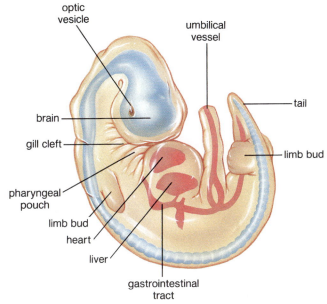

optic
vesicle

umbilical
vessel

brain

tail

gill cleft

limb bud

pharyngeal
pouch

limb bud

heart

liver

gastrointestinal
tract

b.

Figure 9.8

Human embryo at beginning of fifth week. *a.* Scanning electron micrograph. *b.* The embryo is curled so that the head touches the heart, two organs whose development is further along than the rest of the body. The organs of the gastrointestinal tract are forming. The presence of the tail is an evolutionary remnant; its bones will regress and become those of the coccyx. The arms and legs will develop from the bulges that are called limb buds.

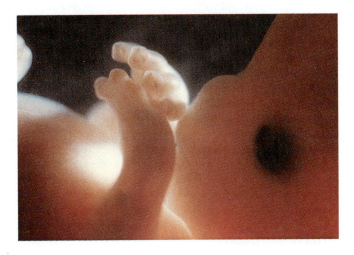

Figure 9.9

Eight-week fetus. The embryonic period is over and from now on a more human appearance takes shape.

HUMAN ISSUE

The cost of caring for a premature baby is astronomical. The smaller a child is at birth, the greater the medical bill (a bill of $150,000 or more is not out of the question) and the more likely the child will be handicapped despite good care. Premature babies often are blind or deaf, have breathing difficulties, are mentally retarded, or have learning disabilities.

Should parents and physicians be allowed to choose whether extreme measures should be taken to save babies below a certain age or birth weight? One physician suggests that only if the parents are very strongly committed to taking full care of their child should measures be taken to save babies whose gestational period is less than 28 weeks and birth weight is below 2.2 pounds. Does this seem reasonable to you?

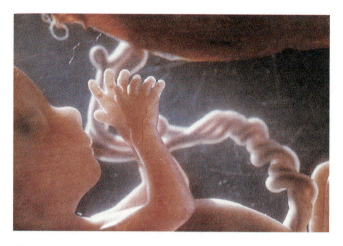

Figure 9.10

Three- to four-month fetus looks human. Face, hands, and fingers are well defined.

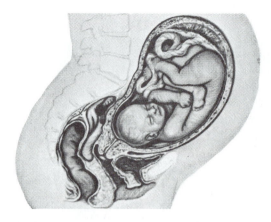

Figure 9.12

Six- to seven-month fetus. By seven months, the uterus reaches above umbilicus of mother, and breasts have enlarged; baby is covered by a fine hair, and eyes are open.

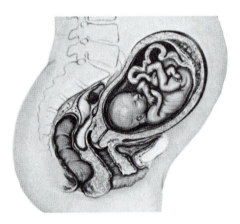

Figure 9.11

Four- to five-month fetus. By five months, the uterus reaches to the umbilicus of mother, and movement is felt; in baby, the skeleton is visible, and heartbeat can be heard.

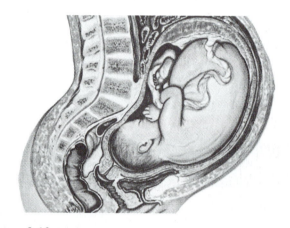

Figure 9.13

Eight- to nine-month fetus. By nine months, the uterus is up to the rib cage of mother, and sleeping is difficult. Baby has gained weight and is ready to be born.

Fetal Development

Table 9.1 lists the main events of fetal development, and figures 9.10 to 9.13 show some of these changes. Of major interest may be the fact that during the third month it would be possible to tell the sex of the **fetus** by visual examination. During the fourth month, the baby is covered by a fine, downlike hair and a bony skeleton has appeared. During the fifth month, expectant mothers can feel the movements of the fetus and the doctor can hear the heartbeat. During the sixth month, the baby is covered by a heavy protective cheesy coating called the vernix caseosa. By the seventh month, premature babies have a fair chance of survival in nurseries staffed by skilled physicians and nurses. Therefore, the baby is said to be viable. During the

eighth and ninth months, the baby puts on weight and the respiratory system completes its development.

Effects on Mother

Table 9.1 also outlines the major changes in the mother during pregnancy. During the first weeks of pregnancy, the mother may experience nausea and vomiting, loss of appetite, and fatigue. Other changes that indicate pregnancy are swelling and tenderness of the breasts, increased urination, and irregular bowel movements. Some women, however, report increased energy levels and a general sense of well-being during this time.

The uterus enlarges greatly during pregnancy. In the nonpregnant female the uterus is only about 2 by 3 inches. Just before giving birth, the uterus almost fills the abdom-

TABLE 9.2	APPROXIMATE WEIGHT GAIN IN PREGNANCY
Full-term baby	7.7 lb
Placenta	1.4 lb
Amniotic fluid	1.8 lb
Enlarged uterus	2.0 lb
Enlarged breasts	0.9 lb
Blood volume increase	4.0 lb
Increased fluid retention	2.7 lb
Maternal storage fat	3.5 lb
	24.0 lb

From H. S. Mitchel, et al., *Nutrition in Health and Disease,* 16th edition. Copyright © 1976 J. B. Lippincott Company, Philadelphia, PA. Reprinted by permission.

inal cavity and reaches even to the rib cage (fig. 9.13). Breasts increase as much as 25% in size. By the end of the sixth month, the nipples and areolae are darkly pigmented, and *colostrum,* a yellow-white fluid that resembles milk but contains more protein and less fat, is being produced.

Toward the end of pregnancy, the enlarged size of the baby and the extra weight cause various difficulties (table 9.2). The mother may have trouble breathing and an increased need to urinate. The center of gravity is thrown forward, therefore standing and walking are difficult. Sleeping may be disturbed not only due to the kicking of the baby but also due to an inability to get comfortable.

BIRTH

The uterus characteristically contracts irregularly (called Braxton Hick's contractions) during the second half of pregnancy. At first, light, often indiscernible contractions last about 20 to 30 seconds, but near the end of pregnancy they become stronger and more frequent so that the woman may falsely think she is in labor. The onset of true labor is marked by uterine contractions that occur regularly every 15 to 20 minutes and last for 40 seconds or more. Birth, also called **parturition** or labor, has three stages. During the first stage, the cervix dilates; during the second, the baby is born; and during the third, the afterbirth is expelled.

The events that cause birth to begin have been investigated for some time, but there are now new findings. It is believed that the fetal hypothalamus directs the fetal pituitary to produce ACTH (adrenocorticotropic hormone), which in turn stimulates the adrenal gland of the fetus to secrete the hormone cortisone. This hormone brings about the production of prostaglandins in the pla-

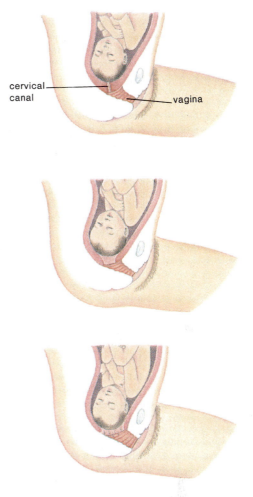

cervical canal

vagina

Figure 9.14

Stage I of parturition: effacement. The cervical canal slowly disappears as baby's head pushes on the lower part of the uterus. This is called effacement or "taking up the cervix."

centa. Prostaglandins are hormones that have a host of effects on the human body, including contraction of the uterus. Prostaglandins may also bring about the release of oxytocin from the maternal pituitary gland. Since oxytocin can cause uterine contractions, either prostaglandins or oxytocin may be given to induce parturition.

Stage I

Prior to the first stage of birth or concomitant with it, there may be a "bloody show" caused by the expulsion of a mucus plug from the cervical canal. This plug prevented bacteria and sperm from entering the uterus during the pregnancy.

Uterine contractions during the first stage of labor occur in such a way that the cervical canal slowly disappears as the lower part of the uterus is pulled upward toward the baby's head (fig. 9.14). This process is called effacement, or "taking up the cervix." With further contractions, the baby's head acts as a wedge to assist cervical dilation. The baby's head usually has a diameter of about 4 inches, therefore, the cervix has to dilate to this diameter in order to allow the head to pass through. If it has not occurred already, the amniotic membrane is apt to rupture now, releasing the amniotic fluid, which escapes out the vagina. The first stage of labor ends once the cervix is completely dilated.

Stage II

During the second stage of birth, uterine contractions occur every 1 to 2 minutes and last about 1 minute each. They are accompanied by the mother's desire to push or bear down. As the baby's head gradually descends into the vagina, the desire to push becomes greater. When the baby's head reaches the vaginal opening it turns so that the back of the head is uppermost (fig. 9.15). Since the vaginal opening may not expand enough to allow passage of the head without tearing, an **episiotomy** is often performed. This incision of the perineum is stitched later and will heal more perfectly than a tear would.

The baby's head emerges from the vagina in degrees—first the forehead, then the nose, and finally the mouth and chin, with the neck situated the entire time just behind the front bone of the pelvic girdle (fig. 9.15). As soon as the head is delivered, the baby's shoulders rotate so that the baby looks either to the right or left. The physician may at this time hold the head and guide it downward while one shoulder and then the other emerges. The rest of the baby follows easily.

The baby's first breath accompanies the events that convert fetal circulation to adult circulation. The change in environmental pressure following birth causes blood to circulate to the lungs rather than entering the arterial duct. The arterial duct contracts and never relaxes. Blood returning from the lungs to the heart exerts pressure against a flap that permanently closes off the oval window. If these circulatory changes do not occur or if there is some other cardiac defect, surgery may be required.

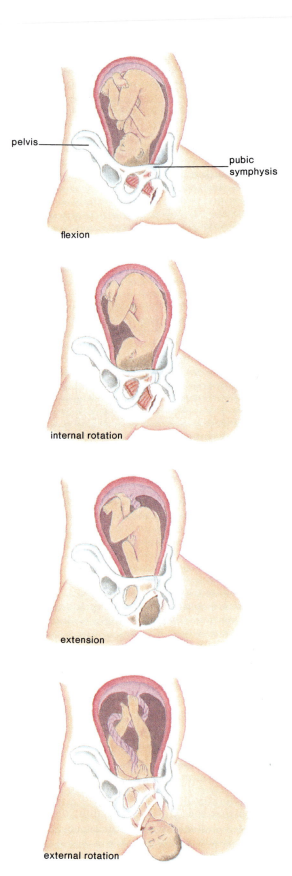

Figure 9.15

Stage II of parturition: delivery of child. The process of birth involves the rotation of the baby's head internally and then externally. These drawings show the woman sitting because a prone position for childbirth is not recommended by many.

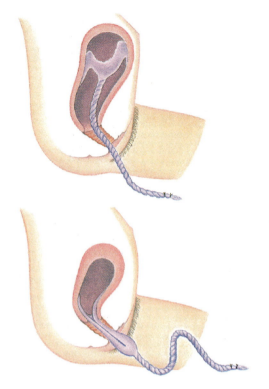

Figure 9.16
Stage III of parturition: delivery of afterbirth. During the last stage of parturition the placenta and amniotic sac are expelled from the vagina.

Once the baby is breathing normally, the umbilical cord is cut and tied, severing the child from the placenta. The stump of the cord shrivels and leaves a scar, which is the navel.

Stage III

The placenta, or afterbirth, is delivered during the third stage of labor (fig. 9.16). About 15 minutes after delivery of the baby, uterine muscular contractions shrink the uterus and dislodge the placenta. The placenta and amniotic sac are then expelled into the vagina. As soon as the placenta and amniotic sac are delivered, the third stage of labor is complete.

Natural Childbirth

Some doctors and expectant couples feel that the nervous system depressants often administered during childbirth may be harmful not only to the expectant mother but to the baby as well. This sentiment, together with a desire to enjoy and share the process of giving birth, gave impetus to the prepared childbirth movement. Couples who wish to practice prepared childbirth using the methods espoused by Dr. Fernand Lamaze and others attend several teaching sessions in which they learn about the events of labor and delivery, the phenomenon of conditioned pain, and suggestions for behavior during labor and delivery.

It is believed that a woman may help prevent discomfort during labor by concentrating on mild, shallow breathing at the time of contractions. This breathing method prevents the diaphragm from exerting pressure on the abdominal organs and guarantees an adequate supply of oxygen for uterine contraction. When delivery begins and the woman feels a great need to push, her partner coaches her to use deep inhalation along with a controlled type of pushing at the time of each strong contraction. Advocates of the Lamaze method of prepared childbirth feel that this active participation on the part of the couple not only helps a woman overlook discomfort but also gives the new parents the pleasant reward of seeing the baby when it first appears.

Lactation

During pregnancy, the breasts enlarge as the ducts and alveoli increase in number and size. The same hormones that affect the mother's breast can also affect the child's. Some newborns, including males, even secrete a little milk for a few days.

Usually, there is no production of milk during pregnancy. The hormone **prolactin** (lactotropic hormone) is needed for lactation to begin, and the production of this hormone is suppressed because of the feedback control that the increased amount of estrogen and progesterone during pregnancy has on the pituitary. Once the baby is delivered, however, the pituitary does begin secreting prolactin. It takes a couple of days for milk production to begin, and, in the meantime, the breasts produce colostrum.

Breast-feeding has come back in vogue in a big way. Not 20 years ago, only one mother in five breast-fed her baby. Today, an estimated two out of every three newborns are being given breast milk. The shift has many advantages, not just for the infant but for the mother as well. One is that when the baby is ready to eat, the milk is ready for the drinking. There are no bottles to sterilize or formulas to measure. Breast milk is also less expensive than formula, even after you account for the cost of the extra food a mother needs to insure that she will produce enough milk and that it will contain enough calories. In addition, hormones released during breast-feeding cause contractions in the uterus that assist it in shrinking closer to its prepregnancy size. And the calories a mother's body spends producing breast milk may help her to lose the 7 to 8 pounds—or more—of fat she puts on during pregnancy but does not "deliver" along with the baby, placenta, and amniotic and other fluids.

As for the baby, "human milk is unquestionably the best source of nutrition . . . during the first months of life," says the American Academy of Pediatrics. That's especially true if conditions in the home are unsanitary (breast milk, unlike formula, does not need to be kept "clean," because it goes directly from mother to child), or money is too scarce to insure that formula will always be affordable, or the educational level of the parents is too low (or the emotional level of the household too "chaotic") for the family to read, fully understand, and consistently apply the rules of proper formula preparation and storage. But even full-term infants born to reasonably well-educated, financially secure parents who live in clean environments may benefit.

It isn't that today's formulas are not reliable. Healthy babies can grow well on breast or bottle. But breast milk is an amazingly sophisticated substance (it contains more zinc in the first few weeks than later on to meet the newborn's higher zinc needs). And although scientists have been able to imitate it safely enough in store-bought formula, "there may still be some subtle, not-yet-discovered ways in which breast milk is preferable to bottle milk," says Ronald Kleinman, M.D., head of the Committee on Nutrition at the American Academy of Pediatrics. "There remain differences between the two," he adds, "whose significance isn't understood."

These differences may be behind such findings as one reported in the *British Medical Journal* that breast-fed infants appear better protected against wheezing during the first few months of life. The journal has also published a report suggesting that breast-fed newborns are less prone to develop stomach and intestinal illness during the first 13 weeks of life and, thus, suffer less than bottle-fed babies from vomiting and diarrhea.

Can the Bottle Ever Be a Better Choice?
Despite the many advantages of breast-feeding, it may not always be the appropriate way for a mother to nourish her child. Some women simply do not want to breast-feed because they are taking prescription drugs that may get into the milk and harm the baby, or cannot breast-feed because they will return to work soon after giving birth and would find "pumping" breast milk to be given to the infant while they are away from home too tiring. Others may not wish to breast-feed because they are uncomfortable with the sexuality of the process or with some other emotionally related aspect. All these reasons are considered valid. In other words, mothers who do not want to breast-feed should not be pressured or made to feel guilty about it. As family therapist and dietitian Ellyn Satter says in her book *Child of Mine: Feeding with Love and Good Sense* (Bull Publishing: Palo Alto, California), "You will have plenty of opportunities to feel guilty as a parent without feeling guilty about *that,* too." Besides, if a mother breast-feeds but hates doing it, her baby is going to sense that, and it will do greater harm than the breast milk will do good.

Indeed, more important than whether a mother breast-feeds, especially in countries like the United States, where the standard of living for the average family is such that breast-feeding is not a life-or-death matter as it is in some developing nations, is that she develops a relaxed, loving relationship with her child. "Even the sophisticated components of breast milk can't make up for that," Ms. Satter says. The closeness, warmth, and stimulation provided by an infant's caretakers are as important to his normal growth and development as the source of his food.

Source: *Tufts University Diet and Nutrition Letter* (ISSN 0747–4105) is published monthly by *Tufts University Diet and Nutrition Letter,* 53 Park Place, New York, NY 10007. This article extracted from a Special Report in the December 1990 issue. Reprinted by permission.

The continued production of milk requires a suckling child. When a breast is suckled, the nerve endings in the areola are stimulated, and a nerve impulse travels from the nipples to the hypothalamus, which directs the pituitary gland to release the hormone oxytocin. When this hormone arrives at the breast, it causes contraction of the lobules so that milk flows into the ducts (called milk letdown), where it may be drawn out of the nipple by the suckling child. The more suckling, the more oxytocin released, and the more milk there is for the child (fig. 9.17). Some women choose to breast-feed and some do not. The reading "Deciding between Feeding by Breast or by Bottle" discusses the benefits of both feeding methods. ■

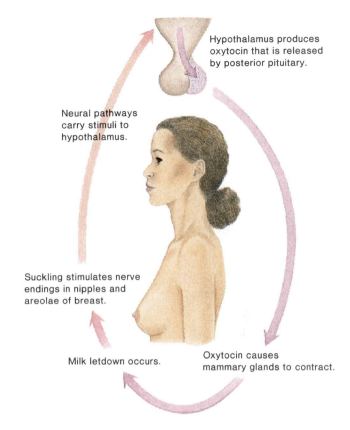

Hypothalamus produces oxytocin that is released by posterior pituitary.

Neural pathways carry stimuli to hypothalamus.

Suckling stimulates nerve endings in nipples and areolae of breast.

Milk letdown occurs.

Oxytocin causes mammary glands to contract.

Figure 9.17

Suckling reflex. Suckling sets in motion the sequence of events that lead to milk letdown, the flow of milk into ducts of the breas (see also fig. 6.13).

SUMMARY

If intercourse and ejaculation occur, sperm may penetrate cervical mucus, particularly at the time of ovulation. During their passage through the female reproductive tract, the sperm undergo capacitation so that acrosome enzymes that allow penetration of the egg are released. Only one sperm head actually enters the egg, and this sperm nucleus fuses with the egg nucleus, which has just undergone the second meiotic division. The zygote begins to develop into an embryo, which travels down the oviduct to imbed itself in the uterine lining. Cells surrounding the embryo

produce HCG, and the presence of this hormone indicates that the female is pregnant.

Human development consists of embryonic and fetal development. During the first 2 months of pregnancy, the extraembryonic membranes appear; they serve important functions, and the embryo acquires organ systems. During the remaining months, the time of fetal development, there is a refinement of these systems.

The fetus, which lies within the amnion surrounded by fluid, is connected to the placenta by means of

the umbilical cord. Birth, or parturition, has three phases. During the first stage, the cervix dilates to allow passage of the baby's head and body. The amnion usually bursts sometime during this stage. During the second stage, the baby is born and the umbilical cord is cut. As the baby takes his or her first breath, anatomical changes convert fetal circulation to adult circulation. During the third stage, the placenta is delivered.

KEY TERMS

allantois 114
amnion 113
chorion 114
congenital defects 114
embryo 116
embryonic development 112

episiotomy 124
extraembryonic membranes 113
fetal development 112
fetus 122
parturition 123

placenta 114
prolactin 125
umbilical cord 116
yolk sac 114
zygote 111

REVIEW QUESTIONS

1. Describe the process of fertilization and the events immediately following.

2. What is the basis of the pregnancy test?

3. Name the four extraembryonic membranes, and give a function for each.

4. Describe the structure and function of the placenta.

5. When during pregnancy does a woman have to be the most careful about the intake of medications and other drugs?

6. Describe the structure and function of the umbilical cord.

7. Specifically, what events normally occur during the first, second, third, and fourth weeks of development? What events normally happen during the second through the ninth months?

8. In general, describe the physical changes in the mother during pregnancy.

9. What are the three stages of birth? Describe the events of each stage.

CRITICAL THINKING QUESTIONS

1. Would you expect a woman who produces a normal amount of estrogen but limited progesterone to menstruate and/or get pregnant? Explain.

2. How can an abdominal pregnancy occur and the fetus come to term and be born?

Birth Control and Infertility

Birth-control methods are used in order to regulate the number of children an individual or couple will have. Birth control is also called contraception, because birth-control methods often are designed to prevent either fertilization or implantation of an embryo in the uterine lining.

BIRTH-CONTROL METHODS

Abstinence is sometimes used as a form of birth control, but our discussion will center on birth-control measures and devices used by sexually active persons. Included are both present means of birth control and future possibilities.

Present Means of Birth Control

One way to assess the effectiveness of present birth-control methods is to indicate the percentage of women who are not expected to get pregnant while using a particular method. For example, with abstinence 100 women out of 100 will not get pregnant. Figure 10.1 divides the means of birth control we will discuss into four categories. The most effective group includes sterilization, the pill, the IUD, and Norplant. The second most effective group includes the diaphragm, cervical cap, vaginal sponge, and condoms. The third most effective group includes coitus interruptus, foam, and nat-

ural family planning (temperature method). The least effective group includes creams, jellies, and natural family planning (calendar method). With the least effective method 75 women out of 100 are not expected to get pregnant while 25 women are expected to get pregnant in the course of a year.

Human blastocyst before implantation.

Sterilization

Sterilization as a means of birth control is a surgical procedure that renders the individual incapable of reproduction. The operations do not affect the secondary sex characteristics of the individual, nor do they necessarily affect sexual performance. Successful vasectomy in the male and tubal ligation in the female allow the individual to engage in sexual activities with no fear of pregnancy.

Vasectomy As the name implies, a **vasectomy** consists of cutting the vasa deferentia (fig. 10.2). The operation is very simple. Two small incisions (sometimes one incision) are made on the scrotum to expose the spermatic cords. The vas deferentia are carefully separated from the other structures in the spermatic cords, and a small section of each is removed. Each end is then sealed so that sperm are unable to travel to the urethra.

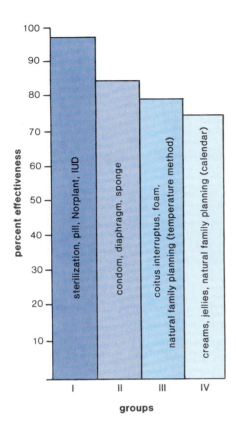

Figure 10.1

Effectiveness of birth-control procedures and devices. Group I is most effective in preventing pregnancy, and group IV is least effective in preventing pregnancy. Source: Data from Alan F. Guttmacher, *Pregnancy, Birth, and Family Planning.* Copyright © 1973 New American Library, New York.

Vasectomies performed on well-adjusted and healthy males do not generally alter sexual motivation, ability to maintain an erection, or ejaculation of semen. However, there is some concern about an unexpected reaction: antibodies against a man's own sperm have been found in the blood following vasectomy. It has been hypothesized that due to the buildup of sperm in the epididymides, sperm are engulfed by white blood cells that later enter the bloodstream. If remnants of the sperm enter the blood from the white blood cells, then antibodies are formed against them. These antibodies will thereafter invade the testes to attack sperm.

Vasectomy should be considered an irreversible operation because resectioning the sperm ducts may be difficult and fertility is usually reduced, perhaps due to an antibody reaction.

Tubal Ligation **Tubal ligation** used to mean simply tying the tubes, but today it refers to an operative procedure in which the oviducts are first cut and then either tied or sealed (fig. 10.3). This prevents sperm from reaching the

egg, because the sperm must halt on the far side of the obstruction.

Laparoscopy, which is a preferred method for tubal ligation, requires only two small incisions that can be covered with adhesive bandages (sometimes called bandaid surgery). First, a local anesthetic is given, and the abdomen is distended with an inert gas in order to give the physician a clear view of the oviducts. A small incision is made near the navel, and the laparoscope (a small telescope with a "cold" light source) is introduced through this cut. A tiny surgical knife inserted through a second incision below the first incision is used to cut and remove a small portion of each tube before they are tied or sealed.

A newer method is called *hysteroscopic sterilization* (hystera means womb). In this procedure, a telescopic device is inserted into the uterus by way of the vagina. This time the oviducts are sealed with an electric current where they enter the uterus. If the operation is successful, scar tissue forms and blocks the tubes. The failure rate, which has been as high as 25%, is reduced if the operation is performed early in the woman's menstrual cycle before the uterine lining has had time to thicken.

Like vasectomy, tubal ligation should be considered an irreversible operation. It is possible to resection the tubes, but this requires abdominal surgery and the resultant fertility is not high.

Birth-Control Pills

The **birth-control pill** is usually a combination of estrogen and progesterone that is taken for 21 days out of a 28-day cycle (fig. 10.4*d*). Either no pill or an inactive pill is then taken for 7 days. The estrogen and progesterone in the pill effectively shut down the pituitary production of both FSH and LH by the feedback mechanism. Theoretically, since the pituitary is not producing FSH, no follicle begins to develop in the ovary, ovulation does not occur, and pregnancy cannot occur. Without the maturation of a follicle, the ovaries do not produce the female sex hormones, but the pill provides these hormones for the patient.

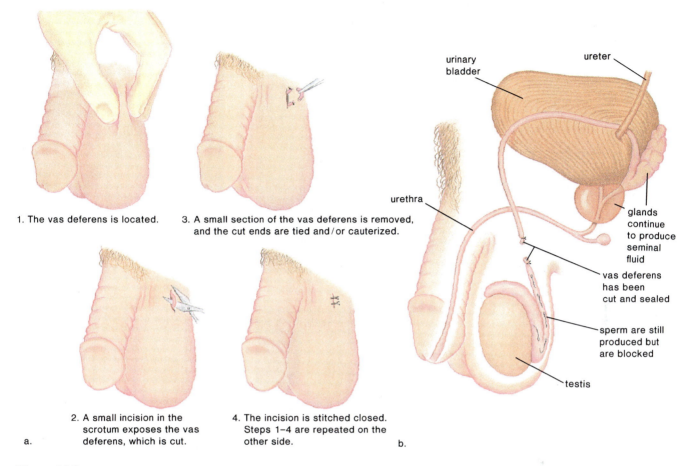

1. The vas deferens is located.

3. A small section of the vas deferens is removed, and the cut ends are tied and/or cauterized.

2. A small incision in the scrotum exposes the vas deferens, which is cut.

4. The incision is stitched closed. Steps 1–4 are repeated on the other side.

a.

urinary bladder

ureter

urethra

glands continue to produce seminal fluid

vas deferens has been cut and sealed

sperm are still produced but are blocked

testis

b.

Figure 10.2

Vasectomy involves cutting and sealing the vas deferentia. *a.* External view of procedure. *b.* Internal view of results.

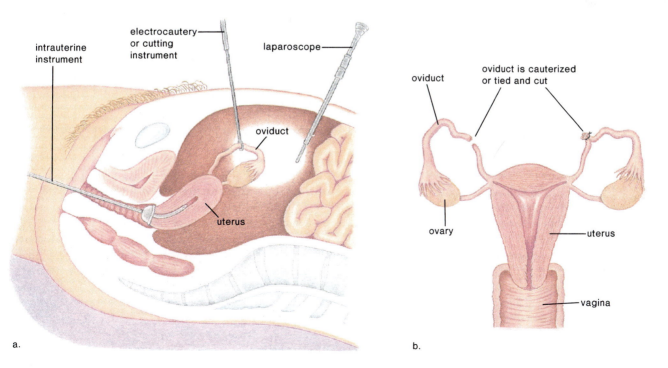

intrauterine instrument

electrocautery or cutting instrument

laparoscope

oviduct

uterus

oviduct

oviduct is cauterized or tied and cut

ovary

uterus

vagina

a.

b.

Figure 10.3

Tubal ligation involves cutting and sealing the oviducts. *a.* Laparoscopy requires only two small incisions. *b.* Internal view of results.

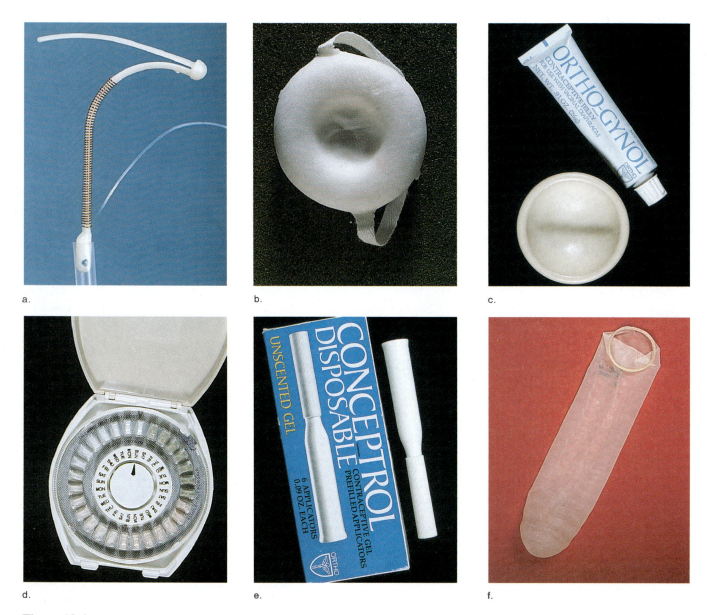

Figure 10.4

Various types of birth-control methods: *a.* IUD; *b.* vaginal sponge; *c.* diaphragm; *d.* birth-control pills; *e.* vaginal spermicide; *f.* condom.

Aside from this primary action of the pill, there are three secondary actions that most likely prevent pregnancy should ovulation happen to occur. The pill prevents the cervical mucus from entering its midcycle phase of being thin and watery. Instead, the mucus remains sticky and fairly impenetrable by sperm. The hormones in the pill could possibly affect the transport of an embryo down the oviducts so that it would not arrive in the uterus at the proper time for implantation. The pill also prevents normal buildup of the lining of the uterus, and therefore an embryo is unable to implant itself. This action of the pill accounts for the fact that menstruation lasts fewer days and the flow is lighter when a woman takes the pill.

Since the pill has one primary action and three secondary actions to prevent pregnancy, it is highly effective. It has been calculated that there should be no more than 1 pregnancy per year per 1,000 women using the pill. However, the pregnancy rate is actually 5 to 7 pregnancies per 1,000 because women do not always follow the directions for use.

Both beneficial and adverse side effects have been linked to the pill. Women report relief of discomforts associated with menstruation and also relief of acne. They also report several minor adverse side effects that are not generally considered to be injurious to health. Many of these are similar to symptoms associated with early pregnancy and are therefore thought to be related to the estrogen in the pill. They include nausea, vomiting, painful breast swelling, and irregular spotting or bleeding. Less common complaints are weight gain, headache, dizziness, and sometimes *chloasma,* which is areas of darkened skin on the face, particularly over the cheekbones. Most often these side effects, except for chloasma, either diminish or disappear by the second or third pill cycle.

A well-documented serious side effect of the pill is an increased incidence of blood clotting within the vessels. Blood clotting normally occurs only when a vessel has been cut, but it can abnormally occur within an intact vessel. If the clot remains stationary and obstructs the flow of blood where formed, it is called a *thrombus.* If the clot is carried to a smaller vessel where it prevents flow, it is called an *embolus.* If the clogged vessel serves a vital organ such as the lung, brain, or heart, serious illness or death may result.

Thromboembolism in women on the pill has been studied extensively in England and the United States. These studies compare the incidence of thromboembolism in two groups of women—one group taking the pill and the other group not taking the pill. The number of women who are hospitalized or who die due to thromboembolism has been shown to be about 5 to 7 times greater in the group taking the pill than in the group not taking the pill. The higher increased incidence occurs when women taking the pill are 35 to 40 years of age and/or are smokers. If the women both smoke and have some other risk factor such as hypertension, the chance of thromboembolism jumps to 78 times as great for a pill user as for a nonuser. It is therefore recommended that these women consider another form of contraception.

People who are against birth control often use these studies to suggest that all women should not take the pill. Proponents of oral contraception compare the risk with that of pregnancy and illegal abortions in order to suggest that oral contraception is safer than these alternatives.

Since there are contraindications to taking the pill, it should always be prescribed by a physician and only after a careful physical examination; since there are side effects, it should be taken only under a physician's continuous care.

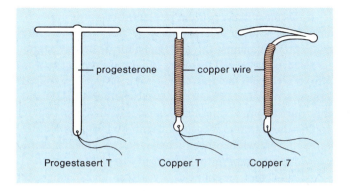

Figure 10.5

Three types of intrauterine devices. Progestasert T has a contraceptive effect because progesterone is embedded in the plastic; copper T and 7 have a contraceptive effect because copper is embedded in the plastic.

Norplant

Norplant is a long-lasting hormonal contraceptive that is implanted under a woman's skin. It consists of six, inch-long, silicone rubber tubes that contain progestin, a synthetic form of progesterone. Progestin leaks through the tubing at a steady rate that maintains blood hormone levels high enough to prevent the hypothalamus from secreting GRH and the anterior pituitary from secreting gonadotropic hormones.

Clinical studies suggest that Norplant has no serious side effects. However, progestin does cause irregular menstrual bleeding in 75% of women who use it. Norplant, which lasts up to 5 years, is said to be 99.7% effective in preventing pregnancy. This would make it the most effective means of reversible contraception available. When the tubes are removed or run out of medicine, fertility is restored in less than 48 hours.

Intrauterine Device (IUD)

An **intrauterine device (IUD)** is a small piece of molded plastic that is inserted into the uterus by a physician (figs. 10.4*a* and 10.5). Two types of IUDs are now available: the copper type has copper wire wrapped around the stem, and the progesterone-releasing type has progesterone embedded in the plastic. The best candidates for an IUD are women who have had at least one child, are of middle to older reproductive age, and who have a stable relationship with a partner who does not have a sexually transmitted disease.

IUDs most likely prevent implantation of the embryo, since there is often an inflammatory reaction where the device presses against the endometrium. Some investigators believe that the IUD alters tubal motility so that the embryo arrives in the uterus before it is properly prepared to receive it. A question arises as to whether an IUD and other birth-control methods cause an abortion. Some people believe that preventing an embryo from implanting is an abortion, while others believe that an abortion is the removal of an embryo that is already implanted.

About 1 to 3 pregnancies a year can be expected per 100 women using an IUD. If pregnancy should occur, it is strongly advised that the device be removed because the chance of a miscarriage is higher if it is left in place (a 50% chance of a miscarriage compared to a 30% chance after removal).

The minor side effects of the IUD are expulsion, pain, irregular bleeding, or profuse menstruation. The device can be expelled from the uterus, especially in women who have not had any children, usually during menstruation and without the woman realizing it. To guard against this, IUDs have an attached nylon thread that projects into the vagina; the wearer can note the presence of the thread by the insertion of a finger.

Perforation of the uterus, that is, puncturing of the wall of the uterus, is the most serious risk of using an IUD. If this occurs, it is almost always at the time of insertion. Perforation must be corrected by surgery because a general infection of the abdominal cavity can result. IUD users have a higher risk of a uterine infection because microorganisms may enter the uterus by "climbing" the thread of the IUD. The infection may spread to the oviducts causing **pelvic inflammatory disease (PID),** with scarring that will increase chances of infertility. The Dalkon Shield was an IUD that had several strands in its tail, and users of this particular IUD showed a 730% higher chance of PID than nonusers. A. H. Robbins Company, the manufacturer of the Dalkon Shield, stopped making the device in 1974 and later filed for bankruptcy because of the expense of defending itself in various lawsuits. As a result, other manufacturers were very reluctant to market IUDs, and only in recent years has the IUD become available in the United States again.

Diaphragm, Cervical Cap, and Vaginal Sponge

The **diaphragm** is a soft rubber or plastic cup with a flexible rim that fits over the cervix (figs. 10.4c and 10.6). A physician determines the proper diaphragm size; therefore, each woman must be individually fitted. The diaphragm must be inserted into the vagina and properly positioned at most 2 hours before sexual relations. It must

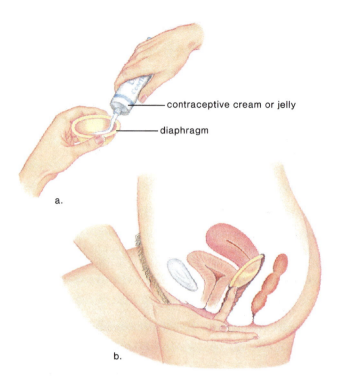

Figure 10.6

Diaphragm. a. Removal from plastic container; b. insertion and position after insertion.

be used with a spermicidal jelly or cream and should be left in place for at least 6 hours after intercourse.

An allergic reaction to the diaphragm or spermicide is the only known side effect. There is a higher rate of pregnancy using the diaphragm (10 to 15 pregnancies per 100 women per year) than with the pill or IUD. This is because insertion may require an interruption to lovemaking, and thus the woman may not bother to use the diaphragm. Some women do not like to handle their genital organs and therefore prefer not to use the diaphragm. However, women can learn to use an inserter to put the diaphragm in place.

The **cervical cap,** a widely used and popular contraceptive device in Europe, is currently being introduced in this country. The cervical cap is thicker and smaller than the diaphragm. The thimble-shaped rubber or plastic cup fits snugly around the cervix. Unlike the diaphragm, the cervical cap is effective even if left in place for several days.

The **vaginal sponge** is shaped to fit the cervix and is permeated with the spermicide nonoxynol-9 (fig. 10.4b). Unlike the diaphragm and cervical cap, the sponge need not be fitted by a physician since one size fits everyone. It is effective immediately after placement in the vagina and remains effective for 24 hours.

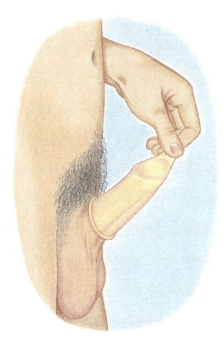

Figure 10.7
Correct method for using a plain-end condom, which has no reservoir for ejaculated semen.

Condom

A **condom** is a thin skin (lambskin) or plastic sheath (latex) that fits over the erect penis (fig. 10.4f). The ejaculate is trapped inside the sheath and therefore does not enter the vagina. When used in conjunction with a spermicidal foam, cream, or jelly, the protection is better than the condom alone, for which 10 to 15 unplanned pregnancies a year per 100 women are expected.

There are no side effects to the condom, and they may be purchased in a drugstore without a prescription. Condoms are checked during the manufacturing process for any possible defects, therefore, it is believed that their relatively high failure rate is due largely to misuse. The condom must be placed on the penis after it is erect, and a small space should be left at the tip to collect the ejaculate (fig. 10.7). If added lubrication is desired, a nonpetroleum-base jelly, such as a spermicidal jelly or cream, should be used since petroleum products—Vaseline, for example—tend to destroy the plastic. Following ejaculation, the upper part of the condom should be held tight against the penis as it is withdrawn from the vagina.

Some men and women feel that using a condom not only interrupts the sex act, but it also dulls sensations; therefore, they prefer one of the other birth-control methods. However, the latex condom is a means of birth control that does offer possible protection against sexually transmitted diseases which will be discussed in the next chapter.

Coitus Interruptus (Withdrawal)

Coitus interruptus is so named because sexual intercourse is abruptly interrupted in order to discharge the semen outside the vagina. Just before orgasm, the male withdraws his penis from the vagina and ejaculates away from the vaginal area. This method of birth control requires careful timing by the male, and he must be sure to direct the penis away from the vagina, since it is possible for sperm deposited near the vagina to work their way into the vagina and up the uterus and to the oviducts.

The advantage of coitus interruptus is that it is always available, but it is not considered good protection, especially since the first drop of seminal fluid, which is released before orgasm, contains numerous sperm. Also, sexual relations are unsatisfactory for some because the male has to concentrate on good timing and the sex act is abruptly discontinued. The rate of unplanned pregnancies is 10 to 20 per 100 women per year with this method of birth control.

Spermicidal Jellies, Creams, and Foams

Jellies, creams, and foams, which contain sperm-killing ingredients such as nonoxynol-9, are inserted into the vagina with an applicator up to 30 minutes before intercourse. A fresh application is required for each subsequent intercourse. Used alone, they are not a highly effective means of birth control, but the foam is more effective because it reaches all parts of the vagina, whereas the jellies and creams tend to localize centrally. It is estimated that 15 to 25 women out of 100 become pregnant each year using this method of birth control.

There are no serious side effects to these products, and they are readily available without prescription in a drugstore. Some women are allergic to certain spermicidal agents and must switch to a new product or discontinue use. Some couples prefer a different method of birth control because of the low protection rate and because insertion might tend to interrupt sexual relations.

Natural Family Planning

Natural family planning, formerly called the rhythm method of birth control, is based on the fact that a woman ovulates only once a month and that the egg and sperm are viable for a limited number of hours or days. The day of ovulation can vary from month to month, and the fertility of the egg and sperm varies perhaps monthly but certainly from person to person. Therefore, the so-called safe period for intercourse without fear of pregnancy cannot be absolutely determined by the use of a calendar alone. However, the following procedure can be utilized by those who are willing to risk an unexpected pregnancy.

1. Keep a record of the length of the monthly cycle for a year. Note the shortest and the longest cycle. A cycle lasts from the first day of menses to its occurrence again.
2. Subtract 18 from the number of days in the shortest cycle. This is the day on which the unsafe period begins.
3. Subtract 11 from the number of days in the longest cycle. This is the day on which the unsafe period ends.

For example, suppose Mary Smith's shortest cycle for the year was 25 days and the longest was 29: $25 - 18 = 7$, and $29 - 11 = 18$. No intercourse should take place between the seventh and nineteenth days of each cycle (fig. 10.8). Using this method, 15 to 30 pregnancies per 100 women per year are expected.

A more reliable way to practice the natural family planning method of birth control is to await the day of ovulation each month and then wait three more days before engaging in intercourse. The day of ovulation can possibly be determined by one of the following methods:

1. Body temperature is lower before ovulation than after it. Immediately preceding ovulation the temperature drops about 0.2°F, but directly following ovulation the temperature rises about 0.6°F (fig. 8.3). Special thermometers, called basal thermometers, are calibrated only from 96°F to 100°F so that the tenths-of-degree marks are widely spaced, making it easier to read the small, expected changes in temperature. The temperature should be taken immediately upon waking in the morning and before leaving bed or starting any activity; the rectal basal temperature is more accurate than the oral basal temperature.
2. The level of sugar in the vagina increases near ovulation. Tes-Tape can be purchased at the drugstore and inserted in the vagina every day.

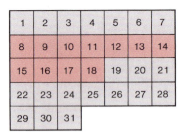

Key:

▨ fertile period—unsafe days

Figure 10.8

This calendar shows the fertile days (unsafe for intercourse) for a woman whose shortest uterine cycle of the year is 25 days and the longest uterine cycle of the year is 29 days. Each individual woman has to work out her own calendar.

When the yellowish tape turns deepest blue, the sugar level is highest and ovulation is near. The level of acidity can also be tested by the use of pH paper. When the paper shows a change from acid to alkaline, ovulation is near.

3. It is possible to test the stringiness and/or the weight of the cervical mucus. When testing the stringiness, the mucus threads are stretched out the longest at the time of ovulation. When testing the weight by means of a special weighing device, the weight decreases at the time of ovulation.

The methods of birth control discussed thus far are the only accepted methods available at the present time. Couples should not rely on other methods suggested by friends, nor should they believe that single events of intercourse will not result in pregnancy. It is obvious that as long as the egg is present and the sperm have been ejaculated into the vagina, a pregnancy is possible.

Future Possibilities for Birth Control

After Intercourse

All of the birth-control methods discussed require planning ahead, even in the case of coitus interruptus, where the male must be prepared to act promptly. It would be convenient to have a method of birth control that could be used after intercourse has actually occurred. Presently, DES, a synthetic estrogen, may be given in large doses following intercourse to prevent pregnancy. This is the same drug that formerly was used to prevent miscarriage in some pregnant women. Unfortunately, it has been shown that the daughters of DES-treated women are more apt to develop a rare type of vaginal cancer. The pregnant women received the drug usually from the sixth week through the twelfth week of pregnancy; the after-intercourse, or postcoital pill, is given for only 5 days.

As a postcoital drug, DES is believed to increase tubular motility so that the embryo arrives early in the uterus, when the lining is not properly prepared. Since DES in the large doses required to prevent pregnancy causes nausea and vomiting, it is not usually administered except in cases of rape and incest.

A birth-control pill (Ru486) on the market in France consists of a synthetic steroid that prevents progesterone from acting on the uterine lining because it has a high affinity for progesterone receptors. In clinical trails the uterine lining sloughed off within 4 days in 85% of women who were less than a month pregnant. To improve the success rate the drug is administered with a small dose of prostaglandin, which causes contraction of the uterus, to expel any embryo present. The promoters of this pill are using the term "contragestation" to describe its effects; however, it should be recognized that this medication, rather than preventing implantation, brings on an abortion, the loss of an implanted fetus. One day the medication might be used by women who experience delayed menstruation but do not know whether or not they are actually pregnant.

Other Possibilities

Various possibilities exist for a "male pill." Scientists have made analogs of gonadotropic-releasing hormones that interfere with the action of this hormone and prevent it from stimulating the pituitary. Experiments in both animal and human subjects suggest that one of these might possibly inhibit spermatogenesis in males (and ovulation in females) without affecting the secondary sex characteristics. The seminiferous tubules produce a hormone termed *inhibin,* which inhibits FSH production by the pituitary. It is hoped that this chemical may someday be produced commercially and made available in pill form for males. Testosterone and/or related chemicals can be injected to

inhibit spermatogenesis in males, but there may be feminizing side effects when the excess is changed to estrogen by the body.

A revival of interest in barrier methods of birth control has occurred, and a "female condom" now is being studied to determine its effectiveness against pregnancy and sexually transmitted diseases (fig. 10.9). The closed end of a large plastic tube is anchored by a plastic ring in the upper vagina, and the open end of the tube is held in place by a thinner ring that rests just outside the vagina.

Abortion

Abortion is the termination of pregnancy before the fetus is capable of survival, which at present is a fetal weight of less than 1 pound. Spontaneous abortions, often called miscarriages, are those that occur naturally, and induced abortions are those that are brought about by external procedures. Although induced abortions are not considered a preferred means of birth control, they are legally available in most states to women who can afford them. The method employed depends on the length of pregnancy.

Uterine Aspiration

Induced abortions during the first 3 months of pregnancy are usually performed by **uterine aspiration.** After a local anesthetic is given, the cervix is dilated and the uterine contents are sucked out by means of a small metal or plastic tube attached to an aspirator (suction machine). The uterus contracts spontaneously, reducing blood loss. Uterine aspiration is sometimes performed when menstruation is 5 to 14 days overdue without definite proof of pregnancy. Under these circumstances, the procedure is called "menstrual regulation" or "menstrual extraction."

Each month K. C. Esperance, 31, a San Francisco nurse practitioner, suffered menstrual cramps so agonizing that she would take to her bed, curl up and pray that she would live through the next couple of days. Doctor after doctor gave her the same ineffectual advice: rest, take some codeine and bear with it.

During her teens, Maria Menna Perper, 42, a New Jersey biochemist, suffered intestinal problems around the time of her period. By her late 30s, she felt "excruciating, burning pain" in her colon every month "like clockwork." Eventually the pain became continuous, and it was impossible for her to work or even sit down.

For Anne Hicks, 29, a Portland, Ore., real estate property manager, there were no obvious signs other than her inability to become pregnant.

Despite their differing complaints, each of the women eventually discovered that she suffered from the same insidious condition: endometriosis, an often unrecognized disease that afflicts anywhere from 4 million to 10 million American women and is a major cause of infertility. The condition is caused by the spread and growth of tissue from the lining of the uterus (or endometrium) beyond the uterine walls. These endometrial cells form bandlike patches and scars throughout the pelvis and around the ovaries and Fallopian tubes, resulting in a variety of symptoms and degrees of discomfort. Because endometriosis has been associated with delayed childbearing, it is sometimes called the "career woman's disease." But recent studies have shown that the disorder strikes women of all socioeconomic groups and even teenagers, though those with heavier, longer, or more frequent periods may be especially susceptible. Says Dr. Donald Chatman of Chicago's Michael Reese Hospital, "Endometriosis is an equal-opportunity disease."

How the disease begins is something of a mystery. One theory ascribes it to "retrograde menstruation." Instead of flowing down through the cervix and vagina, some menstrual blood and tissue back up through the Fallopian tubes and spill out into the pelvic cavity [see figure 10.A]. Normally this errant flow is harmlessly absorbed, but in some cases the stray tissue may implant itself outside the uterus and continue to grow. A second theory suggests that the disease arises from misplaced embryonic cells that have lain scattered around the abdominal cavity since birth. When the monthly hormonal cycles begin at puberty, says Dr. Howard Judd, director of gynecological endocrinology at UCLA Medical Center, "some of these cells get stirred up and could be a major cause of endometriosis."

If anything about endometriosis is clear, it is that once the disease has begun, it will probably get worse. Stimulated by the release of estrogen, the implanted tissue grows and spreads. Cells from the growths break away and are ferried by lymphatic fluid throughout the body, sometimes, although rarely, forming islands in the lungs, kidneys, bowel or even the nasal passages. There they respond to the menstrual cycle, causing monthly bleeding from the rectum or wherever else they have settled.

The most common symptom of endometriosis is pain, which can occur during menstruation, urination, and sexual intercourse. Unfortunately, these warnings are often overlooked by women and their doctors. Cheri Bates, 31, of Seattle, describes the cramps she suffered as "outrageous," but she assumed they were "normal." By the time her condition was discovered, scar tissue covered her reproductive organs and parts of her bladder and intestines.

To confirm that a patient has endometriosis, doctors look for the telltale tissue by peering into the pelvic cavity with a fiber-optic instrument called a laparoscope. After diagnosis, a number of treatments can be prescribed. One is pregnancy—if it is still feasible; the nine-month interruption of menstruation can help shrink misplaced endometrial tissue. Taking birth-control pills may also help, but more effective is a drug called danazol, a synthetic male hormone that stops ovulation and causes endometrial tissue to shrivel. But it also can produce acne, facial-hair growth, weight gain, and other side effects.

A new experimental treatment with perhaps fewer ill effects involves a syn-

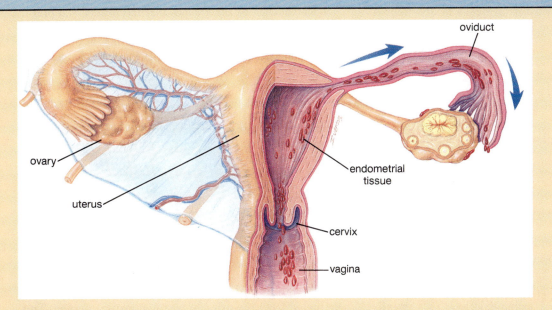

Figure 10.A

Endometriosis. It is speculated that endometriosis is caused by a backward menstrual flow represented by the arrows in this diagram. This allows endometrial cells to enter the abdominal cavity, where they take residence and respond to the monthly cyclic changes in hormonal levels, including those that result in menstruation.

thetic substance called nafarelin, similar to gonadotropin-releasing hormone. Normally GnRH is released in bursts by the hypothalamus gland, eventually triggering the process of ovulation. But "if the GnRH stimulation is given continuously instead of in pulses," explains Dr. Robert Jaffe of the University of California, San Francisco, "the whole [ovulatory] system shuts off," and the endometrial implants "virtually melt away."

For severe cases of endometriosis, surgical removal of the ovaries and uterus may be the only solution. But less extreme surgery can often help. At Atlanta's Northside Hospital, Dr. Camran Nezhat has had success with a high-tech procedure called videolaseroscopy, which employs a laparoscope rigged with a tiny video camera and a laser. The camera images, enlarged on a video screen, enable Nezhat to zero in on endometrial tissue and to vaporize it with the laser. In a study of 102 previously infertile patients, Nezhat found that 60.7% were able to conceive within two years of videolaseroscopy treatment.

Like many other doctors who see the unfortunate consequences of endometriosis, Nezhat is concerned that a "lot of women do not seek help for this problem." Any serious pain, he notes, needs investigating. Agrees Cheri Bates, "If a doctor tells you that suffering is a woman's lot in life, get another doctor."

"The Career Woman's Disease?" in *Time,* April 28, 1986. Copyright © 1986 Time Warner, Inc. Reprinted by permission.

Dilation and Evacuation

During the fourth and fifth months of pregnancy, dilation and evacuation may be used as a means of abortion. A local or general anesthetic is given, and the cervix is dilated by means of a series of cigar-shaped stainless steel dilators of increasing size. Then vacuum suction is used to remove as much as possible of the contents of the uterus. Finally, a long-handled device with a spoon-shaped end, called a curette, is used to scrape the inside of the uterus.

Salt-Induced and Prostaglandin-Induced Abortion

Abortions that are salt induced or prostaglandin induced are used during the sixth month of pregnancy. After this time, the chances of live births are considered to be good, and therefore such abortions are subject to increased legal control.

The most common method in the United States for a late abortion is called salting-out, because it involves the injection of saline (salt solution) or glucose (sugar) solution into the uterus. A quantity of amniotic fluid is removed and replaced by a saline solution. This usually causes the fetus and surrounding membranes to die. Between 5 and 50 hours later, the uterus begins to contract and the fetus is expelled. The uterus may still have to be cleaned by curettage. Complications such as hemorrhaging and infection are more common with late abortions.

In recent years, prostaglandins have been injected into the amniotic fluid to cause abortion by means of induced uterine contractions. The injection-abortion interval has been shortened to less than 20 hours, and there are fewer complications; however, this is a more expensive procedure than salting-out.

INFERTILITY

Sometimes a couple does not need to prevent pregnancy; conception does not occur despite frequent intercourse. The American Medical Association estimates that 15% of all couples in this country are unable to have any children and therefore are properly termed *sterile;* another 10% have fewer children than they wish and therefore are termed *infertile*. The latter category assumes that the couple has been trying to become pregnant and has been unsuccessful for at least 1 year.

Causes of Infertility

The two major causes of **infertility** in females are blocked oviducts, possibly due to pelvic inflammatory disease, discussed in the next chapter, and failure to ovulate due to

There are those who do not approve of alternative methods of reproduction on the grounds that fertilization and pregnancy do not occur in the normal way or that an embryo and, therefore, a human life, might be lost during the procedure. For example, it is possible to freeze IVF embryos for possible use later. What should be done, however, if the mother conceives and does not want the extra embryos, or the couple divorces and the man no longer wishes to father a child? Should unwanted embryos be stored indefinitely and made available to anyone who wishes to have a child? What are the legal rights of frozen embryos, which will eventually die if kept too long?

low body weight. Endometriosis, the spread of uterine tissue beyond the uterus, is also a cause of infertility, as discussed in the reading entitled "Endometriosis." Sometimes these physical defects can be corrected surgically and/or with medication. If no obstruction is apparent and body weight is normal, it is possible to give females HCG extracted from the urine of postmenopausal women. This treatment causes multiple ovulations and sometimes multiple pregnancies.

The most frequent causes of infertility in males are low sperm count and/or a large proportion of abnormal sperm. Disease, radiation, chemical mutagens, too much heat near the testes, and the use of psychoactive drugs can contribute to this condition.

When reproduction does not occur in the usual manner, many couples adopt children. Others sometimes first try one of the alternative reproductive methods discussed following.

Alternative Methods of Reproduction

Artificial Insemination by Donor (AID)

During **artificial insemination,** sperm are placed in the vagina by a physician. Sometimes a woman is artificially inseminated by her husband's sperm. This is especially helpful if the husband has a low sperm count—the sperm can be collected over a period of time and concentrated so that the sperm count is sufficient to result in fertilization. Often, however, a woman is inseminated by sperm acquired from a donor who is a complete stranger to her. At times a mixture of husband and donor sperm are used.

A variation of AID is intrauterine insemination (IUI). IUI involves hormonal stimulation of the ovaries, followed by placement of the donor's sperm in the uterus rather than the vagina.

In Vitro Fertilization (IVF)

In the case of **in vitro fertilization (IVF),** hormonal stimulation of the ovaries is followed by laparoscopy for the purpose of using an aspiratory tube to retrieve preovulatory eggs. Alternately, it is possible to place a needle through the vaginal wall and guide it by the use of ultrasound to the ovaries, where the needle is used to retrieve the eggs. This method is called transvaginal retrieval.

Concentrated sperm from the male are then placed in a solution that approximates the conditions of the female genital tract. When the eggs are introduced, fertilization occurs. The resultant zygotes begin development, and after about 2 to 4 days, the embryos are inserted into the uterus of the woman, who is now in the secretory phase of her menstrual cycle. If implantation is successful, development is normal and continues to term.

Gamete Intrafallopian Transfer (GIFT)

Gamete intrafallopian transfer (GIFT) was devised as a means to overcome the low success rate (15–20%) of in vitro fertilization. The method is exactly the same as in vitro fertilization except the eggs and the sperm are immediately placed in the oviducts after they have been brought together. This procedure is helpful to couples whose eggs and sperm never make it to the oviducts; sometimes the egg gets lost between the ovary and the oviducts, and sometimes the sperm never reach the oviducts. GIFT has an advantage in that it is a one-step procedure for the woman—the eggs are removed and then reintroduced all in the same time period. For this reason, it is less expensive—$1,500 compared with $3,000 and up for in vitro fertilization.

While in most instances of IVF or GIFT, the sperm and eggs have been donated by the prospective parents, this need not be the case. Indeed, IVT or GIFT makes it possible for younger women who have had their ovaries removed or older postpausal women to bear a child. If hormonal therapy is provided, the uterus is still functional in these women.

Surrogate Mothers

In some instances, women have been paid to have babies for individuals who contributed sperm and/or eggs to the fertilization process. If all the alternative methods are considered, it is possible to imagine that a baby could have five parents: (1) sperm donor, (2) egg donor, (3) surrogate mother, and (4) and (5) adoptive mother and father. The courts are in the process of deciding who of these parents have a legal right to the child if there is a dispute. ∎

SUMMARY

Devices and procedures used to prevent conception vary in their effectiveness, and the most effective to the least effective are considered in this chapter. Sterilization, which includes vasectomy in the male and tubal ligation in the female is very effective. The pill, which usually contains both estrogen and progesterone, prevents ovulation due to feedback control over the anterior pituitary. Norplant, which contains progestin, a synthetic progesterone, acts similarly. It is believed that the IUD, an object placed in the uterus, prevents fertilization and/or implantation. The diaphragm, cervical cap, vaginal sponge, and condom, which are barrier methods

of birth control, are somewhat less effective than these. Coitus interruptus, natural family planning, spermicidal foam, jellies, and creams are the least effective of those birth-control measures discussed. Vaginal foam is more effective than jellies and creams. Natural family planning allows intercourse only on "safe" days. Determining these "safe" days is more successful if the woman uses the basal temperature method rather than the calendar method.

Since all these methods have some drawbacks, research continues to find a morning-after, once-a-month, or long-

lasting birth-control procedure. If an unwanted pregnancy does occur, an abortion is sometimes performed. Early abortion procedures, especially uterine aspiration, are preferred over late abortions requiring salting-out or prostaglandin injection.

Some couples are infertile or do not have as many children as they wish. Infertile couples are increasingly resorting to alternative methods of reproduction, including artificial insemination by donor (AID), in vitro fertilization (IVF), gamete intrafallopian transfer (GIFT), and surrogate mothers.

abortion 137

artificial insemination 140

birth-control pill 130

cervical cap 134

coitus interruptus 135

condom 135

diaphragm 134

in vitro fertilization (IVF) 141

infertility 140

intrauterine device (IUD) 133

natural family planning 135

Norplant 133

pelvic inflammatory disease (PID) 134

sterilization 129

thromboembolism 133

tubal ligation 130

uterine aspiration 137

vaginal sponge 134

vasectomy 129

REVIEW QUESTIONS

1. List the birth-control measures discussed and rate the effectiveness of each.

2. Which types of birth-control measures require operative procedures? Why don't these operations affect the secondary sex characteristics?

3. Which birth-control devices must be fitted and/or prescribed by a physician?

4. Which types of birth-control devices may be purchased in a drugstore without a doctor's prescription? Of these, which is the most effective?

5. Theoretically, how does natural family planning work? Explain its high failure rate.

6. Name and discuss future possibilities for birth control. Explain how each works theoretically.

7. Name and discuss the various types of abortions according to the length of pregnancy.

8. What are some treatments for female infertility? for male infertility?

9. What are some alternative reproductive methods?

CRITICAL THINKING QUESTIONS

1. The female birth-control pill contains estrogen and progesterone. A male pill containing the normal amount of testosterone does not seem to work. Why would you have predicted this?

2. Why would you not recommend the production of a birth-control pill for either males or females that contained FSH and LH antagonists?

3. Why would you expect intrauterine insemination to produce better results than the usual method of artificial insemination?

Sexually Transmitted Diseases

Sexually transmitted diseases (STD) are contagious diseases that are passed from one sex partner to another. Most STDs are caused by microorganisms such as bacteria and viruses. Bacteria depend on nutrients supplied by the host, and they often secrete toxins that have a destructive influence on the body's tissues. Viruses are so small that they enter the cells themselves and in this way bring about cellular destruction. The body attempts to ward off infection in two ways: certain white blood cells called **lymphocytes** produce **antibodies** to attack microorganisms, and other white blood cells called neutrophils phagocytize (engulf) them.

In response to some infections, such as the childhood diseases of measles and mumps, adequate antibody production results, and the individual is thereafter immune to an infection. Antibodies are specific, and a new and different type is needed for each infective

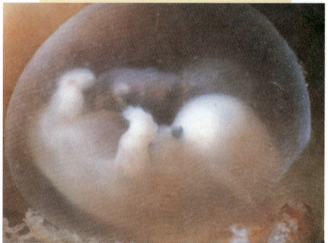

Human embryo at 7 weeks.

agent. For those types of illnesses in which immunity is possible, vaccination is often possible.

Unfortunately, for reasons unknown at this time, humans cannot develop good immunity to any of the STDs. Therefore, prompt and proper medical treatment should be received whenever anyone is exposed to an STD. Persons who find that they have a sexually transmitted disease should tell their sexual contacts so that they can be examined, treated, and counseled. Self-diagnosis and self-treatment are usually inadequate.

The same drugs cannot be used to treat both bacterial and viral infections. **Antibiotics** (e.g., penicillin) are effective against bacteria because they interfere with specific bacterial enzymes. Viruses are not cellular in nature; they consist simply of a nucleic acid core (either DNA or RNA) and a protein coat sometimes covered by a membranous envelope acquired when they leave a host cell. Viruses take over the machinery of the host cell in order to carry out nucleic acid replication and protein synthesis. Antiviral drugs interrupt viral reproduction inside cells by interfering with nucleic acid replication. Interferon is an antiviral protein naturally produced by infected cells in the body and now is mass-produced by recombinant DNA technology (chapter 5).

Figure 11.1 gives the incidence for the most prevalent STDs discussed in this chapter.

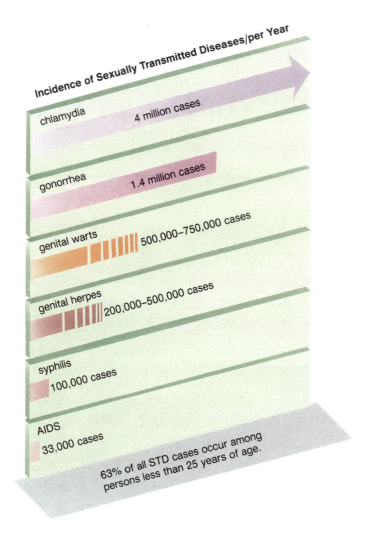

Figure 11.1

Statistics for the most common sexually transmitted diseases. These show that chlamydia, gonorrhea, and genital warts are all much more common than herpes and syphilis. Chlamydia, gonorrhea, and syphilis all can be cured with antibiotic therapy.

STDS OF VIRAL ORIGIN

AIDS

The discussion here may be expanded by the AIDS supplement following this chapter.

The organism that causes **acquired immune deficiency syndrome (AIDS)** is a virus called **human immunodeficiency virus (HIV).** HIV attacks a type of lymphocyte known as **helper T cells.** Helper T cells stimulate the activities of B lymphocytes, which produce antibodies. After an HIV infection sets in, helper T cells begin to decline in number, and the person becomes more susceptible to other types of infections. In other words, the HIV virus brings about destruction of lymphocytes that help protect us from diseases.

Symptoms

AIDS has three stages of infection. During the first stage, which may last about a year, the individual is an asymptomatic carrier. The AIDS blood test (an antibody test) is positive; the individual can pass on the infection, yet there are no symptoms. During the second stage, called **AIDS-related complex (ARC),** which may last 6 to 8 years, the lymph glands are swollen and there may also be weight loss, night sweats, fatigue, fever, and diarrhea. Infections like thrush (white sores on the tongue and in the mouth) and herpes (discussed following) recur. Finally, the person may develop full-blown AIDS, especially characterized by the development of an opportunistic disease such as an unusual type of pneumonia, skin cancer, and also nervous disorders. Opportunistic diseases are ones that occur only in individuals who have little or no capability of fighting an infection. The AIDS patient usually dies about 7 to 9 years after infection.

Transmission

AIDS is transmitted by infected blood, semen, and vaginal secretions. In the United States the two main affected groups are homosexual men and intravenous drug abusers (and their sexual partners). In Africa and some parts of South America, though, AIDS is apparently transmitted chiefly through heterosexual intercourse, and an equal number of men and women are infected.

Certain portions of the United States have been harder hit than others. Even in New York City, which reports the highest level of infectivity, there are areas that report more cases than others. The AIDS virus can cross the placenta, and the Bronx reports 1 in 43 babies are born with HIV antibodies in the blood. Some of these newborns may have the antibodies without having the virus, but 30% to 50% are most likely infected.

Although intravenous drug users serve as a way to spread the disease to the general heterosexual population, to date infection among the general population is about 4%, increasing less than 1% from 1985 to 1990. Health officials emphasize that unprotected intercourse with multiple partners or a single infected partner increases the chance of transmission. The use of a latex condom along with a spermicide containing nonoxynol-9 reduces the risk, but the very best preventive measure at this time is a long-term mutually monogamous relationship with a sexual partner who is free of the disease. Casual contact with someone who is infected, such as shaking hands, eating at the same table, or swimming in the same pool, is not a mode of transmission.

Treatment

The drug AZT has been found to be helpful in prolonging the lives of AIDS patients. Another drug, DDI, which like AZT works by preventing viral replication in cells, is undergoing clinical trials.

Investigators are trying to develop a vaccine for AIDS. The AIDS virus mutates frequently, but researchers have identified a portion of the coat envelope that they believe is relatively stable. When this is injected into the blood-stream antibodies do develop, but it is not yet known whether such antibodies will offer protection against infection. After all, persons with AIDS do produce antibodies for several years, but for some reason the number of helper T cells still declines. Then antibody production falls off.

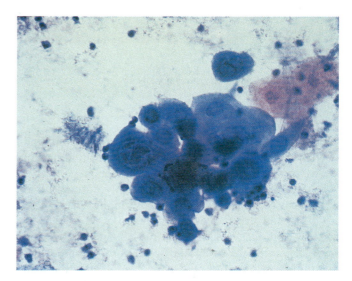

Figure 11.2
Cell infected with herpes virus.

Genital Herpes

Genital herpes is caused by the *Herpes simplex* **virus (HSV),** of which there are two types: type 1 usually causes cold sores and fever blisters, while type 2 more often causes genital herpes (fig. 11.2). Crossover infections do occur, however.

Transmission and Symptoms

An estimated 40 million persons (17% of the United States population) presently have genital herpes; an estimated 500,000 new cases appearing each year. Only a few infections ever seem to cause recognizable symptoms. Immediately after infection, the individual may experience a tingling or itching sensation before blisters appear on the genitals within 2 to 20 days (fig. 11.3). Once the blisters rupture, they leave painful ulcers that may take as long as 3 weeks or as little as 5 days to heal. The ulcers

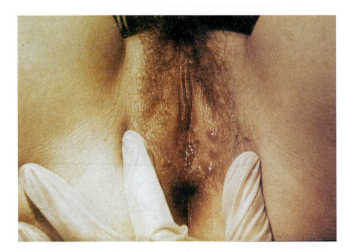

Figure 11.3

A *Herpes simplex* infection sometimes causes blisters on the genitals, but often there are no symptoms at all.

may be accompanied by fever, pain upon urination, and swollen lymph nodes. At this time the individual has an increased risk of acquiring an AIDS infection.

After ulcers heal, the disease is only dormant, and blisters can recur at variable intervals. Sunlight, sex, menstruation, and stress seem to cause the symptoms of genital herpes to recur. While the virus is latent it resides in nerve cells near the brain and spiral cord. Herpes occasionally infects the eye, causing an eye infection that can lead to blindness, and it can also cause infections of the central nervous system. Genital herpes formerly was thought to cause a form of cervical cancer, but this is no longer believed to be the case.

Infection of the newborn can occur if the child comes in contact with a lesion in the birth canal. In 1 to 3 weeks, the infant is gravely ill and can become blind, have neurological disorders including brain damage, or die. Birth by cesarean section prevents these occurrences.

Treatment

Presently there is no cure for herpes. The drugs vidarabine and acyclovir disrupt viral reproduction. The ointment form of acyclovir relieves initial symptoms, and the oral form prevents the occurrence of outbreaks. Work is also being done to develop a vaccine for herpes.

Genital Warts

Genital warts are caused by the human papilloma viruses (HPVs), which are viruses that reproduce in the nuclei of skin cells. Plantar warts and common warts are also caused by HPVs.

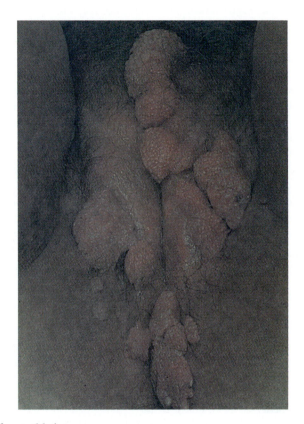

Figure 11.4

A papilloma virus infection sometimes causes genital warts on the genitals, but often there are no symptoms at all.

Transmission and Symptoms

At the moment, genital and anal HPV infections appear to be the most prevalent STDs in the United States. Many times carriers do not have any sign of warts, although flat lesions may be present. When present, the warts commonly are seen on the penis and foreskin of males and the vaginal opening in females (fig. 11.4). If the warts are removed, they may recur.

HPVs, rather than genital herpes, now are associated with cancer of the cervix, as well as tumors of the vulva, the vagina, the anus, and the penis. Some researchers believe that the viruses are involved in 90% to 95% of all cases of cancer of the cervix. Physicians are disheartened that teenagers with multiple sex partners seem to be particularly susceptible to HPV infections. More and more cases of cancer of the cervix are being seen among this age group.

Treatment

Presently, there is no cure for an HPV infection, but warts can be treated effectively by surgery, freezing, application of an acid, or laser burning, depending on severity. A suitable medication to treat genital warts before cancer occurs is being sought, and efforts also are underway to develop a vaccine.

STDS OF BACTERIAL ORIGIN

Gonorrhea

Gonorrhea is caused by the bacterium *Neisseria gonorrheae,* which is a diplococcus, meaning that generally two spherical cells stay together. Although the incidence of gonorrhea declined overall during the 1980s, the incidence among black men and women has increased.

Symptoms

The diagnosis of gonorrhea in the male is not difficult, as long as he displays typical symptoms (as many as 20% of males may be asymptomatic). The patient complains of pain on urination and has a thick, greenish yellow urethral discharge 3 to 5 days after contact (fig. 11.5). In the female, the bacteria may first settle within the urethra or near the cervix, from which they may spread to the oviducts, causing **pelvic inflammatory disease (PID).** PID from gonorrhea is especially apt to occur in the female using an intrauterine device as a birth-control measure (chapter 10). As the inflamed tubes heal, they may become partially or completely blocked by scar tissue. As a result, the female is sterile or at best subject to ectopic pregnancy. Unfortunately, 60% to 80% of females are asymptomatic until they develop severe pains in the abdominal region due to PID. PID affects about 1 million women a year in the United States. Similar to females, in untreated males

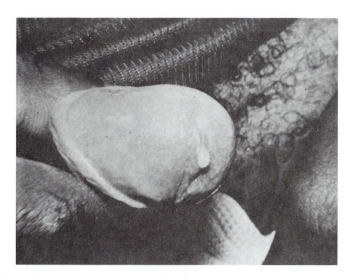

Figure 11.5
Gonorrhea in the male can be recognized by a greenish yellow discharge.

there may be inflammation followed by scarring of the vas deferens.

Homosexual males develop gonorrhea proctitis, or infection of the anus, with symptoms including pain in the anus and blood or pus in the feces. Oral sex can cause infection of the throat and the tonsils. Gonorrhea also can spread to other parts of the body, causing heart damage or arthritis. If, by chance, the person touches infected genitals and then his or her eyes, a severe eye infection can result (fig. 11.6). Eye infection leading to blindness can occur as a baby passes through the birth canal. Because of this, all newborn infants receive eyedrops containing antibacterial agents, such as silver nitrate, tetracycline, or penicillin, as a protective measure.

Transmission and Treatment

Gonococci live for a very short time outside the body; therefore, most infections are spread by intimate contact, usually sexual intercourse. A female has a 50% to 60% risk of contracting the disease after a single exposure to an infected male, whereas a male has a 20% risk after exposure to an infected female.

Unfortunately, there is no blood test for gonorrhea; it is necessary to culture the patient's discharge to identify the organism positively. This fact makes it very difficult to identify asymptomatic carriers who are capable of

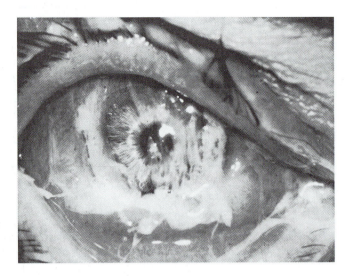

Figure 11.6
Gonorrhea infection of the eye is possible whenever the bacteria comes in contact with the eyes. This can happen when the newborn passes down the birth canal. Manual transfer from the genitals to the eyes is also possible.

passing on the condition without realizing they are doing so. However, if the infection is diagnosed, gonorrhea can be treated using antibiotics. There is no vaccine for gonorrhea, and immunity does not seem possible. Therefore, it is possible to contract the disease many times over.

Chlamydia

Chlamydia is named for the tiny bacterium that causes it (*Chlamydia trachomatis*). New chlamydial infections occur at an even faster rate than gonorrheal infections. They are the most common cause of **nongonococcal urethritis (NGU).** About 8 to 21 days after infection, men experience a mild burning sensation upon urination and a mucoid discharge. Women may have a vaginal discharge along with the symptoms of a urinary tract infection. Unfortunately, a physician mistakenly may diagnose a gonorrheal or urinary infection and prescribe the wrong type of antibiotic, or the person may never seek medical help. In either case, the infection can cause PID and sterility or ectopic pregnancy. ·

If a newborn comes in contact with chlamydia during delivery, inflammation of the eyes or pneumonia can result. There are also those who believe that chlamydial infections increase the possibility of premature and stillborn births.

New and faster laboratory tests are now available for chlamydia detection. Their expense sometimes prevents public clinics from using them, however. It has been suggested that the following criteria could help physicians: women should be tested who (1) are no more than 24 years old; (2) have had a new sex partner within the preceding 2 months; (3) have a cervical discharge; (4) bleed during parts of the vaginal exam; and (5) use a nonbarrier method of contraception. Some doctors, however, are routinely prescribing additional antibiotics appropriate to treating chlamydia for anyone who has gonorrhea, because 40% of females and 20% of males with gonorrhea also have chlamydia.

Chlamydia causes cervical ulcerations that increase the risk of acquiring an HIV infection. Use of a condom serves as a protection against spreading these STDs. The concomitant use of a spermicide containing nonoxynol-9 gives added protection.

Syphilis

Syphilis is caused by a bacterium called *Treponema pallidum.* The disease has three stages, which can be separated by latent periods in which the bacteria are resting before multiplying again. During the primary stage, a hard **chancre** (ulcerated sore with hard edges) indicates the site of infection (fig. 11.7*a*). The chancre can go unnoticed, especially since it usually heals spontaneously, leaving little scarring. During the secondary stage, proof that bacteria have invaded and spread throughout the body is evident when the victim breaks out in a rash (fig. 11.7*b*). Curiously, the rash does not itch and is seen even on the palms of the hands and the soles of the feet. There can be hair loss and infectious gray patches on the mucous membranes, including the mouth. These symptoms disappear of their own accord.

During the tertiary stage, which lasts until the patient dies, syphilis may affect the cardiovascular system; weakened arterial walls (aneurysms) are seen, particularly in the aorta. In other instances, the disease may affect the nervous system; an infected person may become mentally retarded, blind, or walk with a shuffle or show signs of insanity. **Gummas,** which are large destructive ulcers, may develop on the skin or within the internal organs (fig. 11.7*c*).

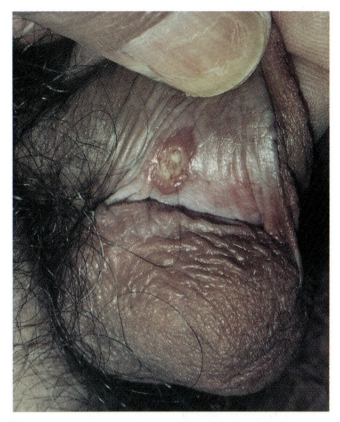

a.

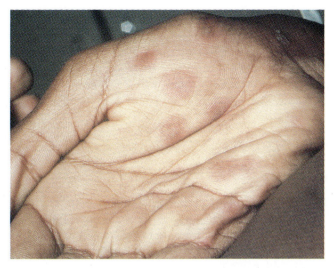

b.

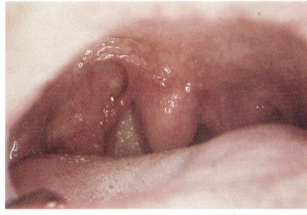

c.

Figure 11.7

Stages of syphilis. *a.* During the primary stage a chancre appears. *b.* During the secondary stage a rash appears on the body, including the palms of hands and soles of feet. *c.* During the tertiary stage there may be gummas—large open sores like this one is on the roof of the mouth.

Congenital syphilis is caused by syphilitic bacteria crossing the placenta. The child is stillborn or blind with many other possible anatomical malformations.

The syphilis bacterium is not present in the environment; it is only present within human beings. Only close intimate contact, such as sexual intercourse, transmits the condition from one person to another.

Diagnosis of syphilis can be made by dark-field microscopic examination of fluids from lesions for the bacterium, which is actively motile and has a corkscrew shape. Blood tests can also detect the bacterium, but they are not positive until at least 6 weeks after initial infection.

Syphilis is a very devastating disease. Controlling it depends on prompt and adequate treatment of all new cases; therefore, it is very important for all sexual contacts to be traced so that they can be treated. Use of condoms can prevent exposure. If syphilis is diagnosed, it can be treated using antibiotics. The incidence of syphilis is rising in most parts of the world; 100,000 cases are reported annually (fig. 11.1), and the highest incidence of these cases is among persons 29 to 39 years of age.

Chancroid

Chancroid is caused by the bacterium *Haemophilus ducreyi* that requires the special growth factors found only in blood and other body fluids.

This STD begins with the development of a small pustule at the point of infection, usually the genitals. The pustule ruptures and forms an ulcer similar to the chancre observed with syphilis. The ulcer is often accompanied by swollen and oozing lymph nodes usually within the groin. Specific diagnosis is made by isolating the bacterium from a lesion or lymph node. Antibiotic therapy is expected to be effective but resistant strains do exist.

Between the end of World War II and up until 1984, chancroid was rarely seen in the United States. Beginning in 1984 a series of outbreaks have occured in major cities. The total number of cases rose from 665 in 1984 to 4,714 in 1989. Unfortunately those with this infection have an increased risk of acquiring an AIDS infection.

STDS OF OTHER ORIGINS

Vaginitis

Two other types of organisms are of interest when studying STDs: protozoans and fungi. Protozoans, which are one-celled organisms that differ as to their mode of locomotion, are usually harmless. They live an independent existence and move by extensions of the cytoplasm; some move by cilia, and still others have flagella. Protozoans are most often found in an aquatic environment, such as freshwater ponds, and the ocean teems with them. All protozoans require an outside source of nutrients, but only the parasitic ones take this nourishment from their host.

Similarly, most fungi are nonpathogenic. Usually the body of a fungus is made up of filaments called hyphae, but yeasts are an exception since they are single cells. Almost everyone is familiar with yeast; because of its ability to ferment, it is used to make bread rise and to produce wine, beer, and whiskey.

Females very often have vaginitis, or infection of the vagina, that is caused by the flagellated protozoan *Trichomoniasis vaginalis* or the yeast *Candida albicans* (fig. 11.8). The protozoan infection causes a frothy white or yellow foul-smelling discharge accompanied by itching, and the yeast infection causes a thick, white, curdy discharge, also accompanied by itching. **Trichomoniasis** is most often acquired by sexual intercourse, and the asymptomatic male is usually the reservoir of infection. *Candida albicans,* however, is a normal organism found in the vagina; its growth simply increases beyond normal under certain circumstances. Women taking birth-control pills are sometimes prone to yeast infections, for example. Also, the indiscriminate use of antibiotics can so alter the normal balance of organisms in the vagina that yeast infections will flare up.

Pubic Lice (Crabs)

Small lice, which are animals that look like crabs under low magnification, infect the pubic hair, underarm hair,

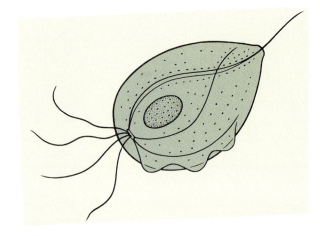

Figure 11.8

Trichomonas vaginalis. The protozoan that causes vaginitis is pear-shaped and uses flagella to move about.

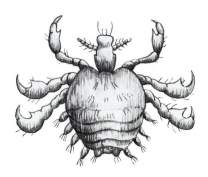

Figure 11.9

Phthirus pubis, the parasitic louse that infects the pubic hair of humans.

and even occasionally the eyebrows of humans. The females lay their eggs around the base of the hair, and these eggs hatch within a few days to produce a larger number of animals that suck blood from their host and cause severe itching, particularly at night.

The crab louse can be contracted by direct contact with an infected person or by contact with his or her clothing or bedding (fig. 11.9). In contrast to most types of sexually transmitted diseases, self-treatment (that does not require shaving) is possible. The lice can be killed by gamma benzene hexachloride, marketed in shampoo, lotion, or cream under the tradename *Kwell.* ■

SUMMARY

Certain STDs are caused by viruses. AIDS is caused by HIV (human immunodeficiency virus), which attacks a type of white blood cell known as helper T-cells. Since helper T-cells stimulate the B cells, which produce antibodies against disease, AIDS is characterized by reduced immunity to diseases. There are three stages. During the first stage (6 to 12 months), the individual is a symptomatic carrier and the virus can be passed on even though the individual has no symptoms. During the second stage (1 to 6 years), there may be only swollen lymph nodes; then weight loss, fatigue, and diarrhea occur before a skin disease like thrush or herpes appears. Finally, opportunistic diseases, such as an unusual form of pneumonia or skin cancer, and mental disturbances occur. In the United States the disease is most prevalent among homosexual men and intravenous drug users. Drug treatment is available, and various types of vaccines are being investigated. Genital herpes, characterized by painful blisters, is caused by the herpes simplex virus. Genital warts are caused by the human papilloma virus. Asymptomatic carriers for both of these viral infections are common. A papilloma virus infection is associated with the occurence of cervical cancer.

Other STDs are caused by bacteria. Gonorrhea is more easily detected in the male because of painful urination; the female is often asymptomatic. Eventual sterility due to PID is possible in both sexes. Chlamydia is caused by a tiny bacterium, and the symptoms are similar to a urinary infection. PID also occurs with chlamydia. Chlamydia and genital herpes infections cause an increased risk of acquiring AIDS. Syphilis, which has three stages (chancre, rash, gummas) can end in cardiovascular and brain impairment. Other STDs are caused by protozoans, fungi, and small lice that look like crabs.

Fetuses can be infected with AIDS and syphilis while still in the womb. Other STDs are acquired as the newborn passes through the vagina.

KEY TERMS

acquired immune deficiency syndrome (AIDS) 144

AIDS-related complex (ARC) 144

antibiotics 143

antibodies 143

Candida albicans 150

chancre 148

chancroid 149

chlamydia 148

congenital syphilis 149

genital herpes 145

genital warts 146

gonorrhea 147

gummas 148

helper T cells 144

herpes simplex virus 145

human immunodeficiency virus (HIV) 144

lymphocytes 143

Neisseria gonorrheae 147

nongonococcal urethritis (NGU) 148

pelvic inflammatory disease (PID) 147

sexually transmitted diseases (STD) 143

syphilis 148

Treponema pallidum 148

trichomoniasis 150

Trichomoniasis vaginalis 150

REVIEW QUESTIONS

1. What is the normal method by which the body fights infection? Are these methods effective against sexually transmitted diseases?

2. Name the cause, symptoms, and mode of transmission for AIDS. What two groups of people have the highest incidence?

3. What two infections of the genitals are caused by viruses?

4. Why might a physician confuse the symptoms of chlamydia with those of gonorrhea? What is the end result of PID?

5. What are the three stages of syphilis? What causes congenital syphilis?

6. What are the symptoms of a *Trichomonas* infection? of a *Candida albicans* infection?

7. Crabs is caused by what type of animal? How do you get crabs?

CRITICAL THINKING QUESTIONS

1. Why wouldn't penicillin, an antibiotic that cures syphilis, be effective against AIDS? And why wouldn't a physician prescribe AZT for syphilis?

2. Relate the symptoms of AIDS, genital herpes, genital warts, gonorrhea, chlamydia, and syphilis to the activities of the causative agent.

3. Both HIV and *Treponema pallidum* lead to illness throughout the body. Discuss similarities and differences between the two infections.

FURTHER READINGS FOR PART II

Aral, S. O., and K. K. Holmes. February 1991. Sexually transmitted diseases in the AIDS era. *Scientific American.*

Baconsfield, P., et al. August 1980. The placenta. *Scientific American.*

The Boston Women's Health Book Collective. 1976. *Our bodies, ourselves: A book by and for women,* 2d ed. New York: Simon & Schuster.

Crapo, L. 1985. *Hormones: The messengers of life.* New York: W. H. Freeman & Co.

Deamer, D. W. 1981. *Being human.* Philadelphia: CBS College Publishing.

DeWitt, W. 1989. *Human biology.* Glenview, Ill.: Scott, Foresman and Co.

Goldberg, S., and B. DeVitto. 1983. *Born too soon: Preterm birth and early development.* San Francisco: W. H. Freeman & Co.

Guttmacher, A. F., and I. H. Kaiser. 1987. *Pregnancy, birth and family planning.* New York: Viking Press.

Katchadourian, H. A., and D. T. Lunde. 1980. *Fundamentals of human sexuality,* 3d ed. New York: Holt, Rinehart & Winston.

Lein, A. 1979. *The cycling female. Her menstrual rhythm.* San Francisco: W. H. Freeman & Co.

Mills, J. and H. Masur. August 1990. AIDS-related infections. *Scientific American.*

Neufeld, P. J. and N. Colman. May 1990. When science takes the witness stand. *Scientific American.*

Nilsson, L. 1977. *A child is born.* New York: Delacorte Press.

Oldstone, Michael B. A. August 1989. Viral alteration of cell function. *Scientific American.*

Rugh, R., and L. B. Shettles. 1971. *From conception to birth: The drama of life's beginnings.* New York: Harper & Row.

Scientific American. October 1988. Entire issue is devoted to articles on AIDS.

Ulmann, A., G. Teutsch, and D. Philibert. June 1990. RU 486. *Scientific American.*

Vander, A. J., et al. 1985. *Human physiology,* 4th ed. New York: McGraw-Hill.

Volpe, E. 1983. *Biology and human concerns.* Dubuque, Ia: Wm. C. Brown.

Wassarman, P. M. December 1988. Fertilization in mammals. *Scientific American.*

AIDS Supplement

Acquired immune deficiency syndrome (AIDS) is caused by a group of related viruses known as HIV (human immunodeficiency viruses). The full name of AIDS can be explained in this way: *acquired* means that the condition is caught rather than inherited; *immune deficiency* means that the virus attacks the immune system, so those infected become more susceptible to infectious diseases and cancer; and *syndrome* means that some fairly typical infections and cancers occur in infected persons.

Presently there is no cure for AIDS, and the number of diagnosed cases is increasing (fig. A.1). AIDS is not distributed equally throughout the country: at the present time, New York City, San Francisco, and Los Angeles alone have over one-third the cases. However, AIDS is now spreading throughout the country.

Although over half of AIDS patients are white, blacks and hispanics have a higher proportionate number of cases compared to whites. Presently, 90% of all AIDS patients are males, and only 10% are women. These statistics are understandable when one realizes that homosexual men are a high-risk group for AIDS; the other high-risk group is IV (intravenous) drug users. The behavior of these two groups places them at risk because of the mode of transmission of HIV.

TRANSMISSION OF AIDS

Human immunodeficiency viruses infect the blood. Viruses do not have the capability of independent existence; they must live and reproduce inside a cell. The original host cell dies as the new viruses depart to infect other cells. Also, the immune system has the capability of destroying infected cells. HIV viruses live and reproduce inside a type of white blood cell called a helper T lymphocyte, sometimes shortened to T4 cell. Lymphocytes are the type of white blood cell that allows us to become immune to diseases and protects us from cancer. There are two types of lymphocytes—the B cells and the T cells. When a person is exposed to an infectious microbe, the B cells are stimulated by helper T cells to produce specific antibodies, which are molecules that can combine with and inactivate a specific microbe by combining with it. Helper T cells also activate killer T cells that attack cancer cells directly. After HIV viruses have infected the blood, the number of helper T cells (T4 cells) eventually decreases, and the individual becomes susceptible to more and more diseases.

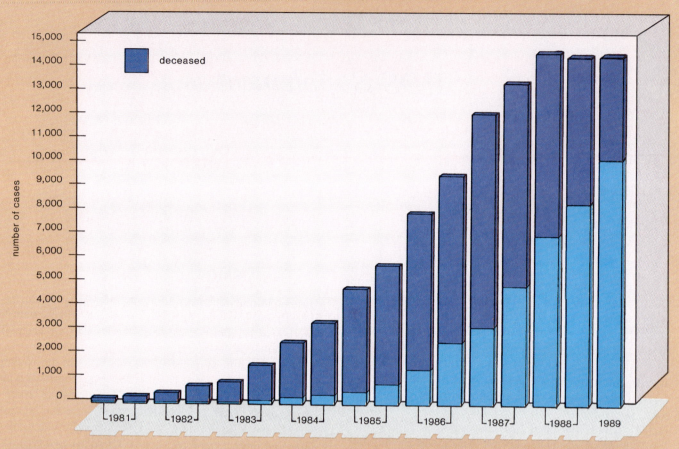

Figure A.1
Reported cases of AIDS in the United States since 1981. For the total number of people who have contracted AIDS, add up all those reported per each half year; for the total number of people who have died, add up all those represented by the dark color. Most people die within several years of contracting AIDS. Sources: Data from W. L. Heyward and J. W. Curran, "The Epidemiology of AIDS in the U.S.," *Scientific American*, October 1988; and *HIV/AIDS Surveillance*, Year-End Edition, January 1990, U.S. Department of Health and Human Services.

IV drug users are a high-risk group for AIDS because HIV can be transmitted if they share contaminated needles. When an IV drug user is infected with HIV, the needle becomes contaminated with the virus and the next person using that needle becomes infected also. About 28% of total cases of AIDS are IV drug users (fig. A.2).

A small number of persons (about 3% of total cases) have contracted AIDS by receiving infected blood during a blood transfusion. Particularly, hemophiliacs, who require more frequent blood transfusion than most or routine injections of a blood product called factor-VIII, have been known to get AIDS in this manner. Since 1985, blood donations in the United States have been tested for contamination. For various reasons, the testing process is not absolutely perfect, and therefore the slight possibility remains of getting AIDS from a blood transfusion. Persons

who are about to undergo an operation can predonate their own blood and/or have noninfected friends do so.

AIDS is a sexually transmitted disease. Semen can contain the virus or more likely infected lymphocytes. Anal-rectal intercourse is common among homosexual men, who account for 60% of the total cases. The lining of the rectum is a thin, single cell layer that is more easily torn during intercourse. Viruses or infected lymphocytes within semen deposited in the rectum can therefore pass into the blood. The vagina has a thicker layer of cells than the rectum, and it does not as easily allow the passage of the virus or infected lymphocyte into the blood. The uterus is a different matter, however. Its thinner and more vulnerable walls do allow the passage of a virus or infected lymphocyte into the blood. Once a female is infected, her vaginal secretions might very well contain the virus and/or infected lymphocytes.

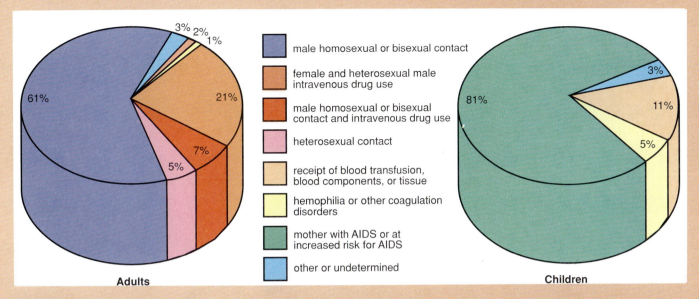

Figure A.2

The distribution of AIDS among adults and children. Homosexual or bisexual men and IV drug users account for 89% of all adult cases. Most children acquired the disease from a mother who was a carrier or had AIDS. Sources: Data from W. L. Heyward and J. W. Curran, "The Epidemiology of AIDS in the U.S.," *Scientific American,* October 1988; and *HIV/AIDS Surveillance,* Year-End Edition, January 1990, U.S. Department of Health and Human Services.

The use of latex condoms can help prevent the spread of AIDS. If infected semen is trapped within the condom, the virus and/or infected lymphocyte obviously cannot pass into the blood and infect a partner. Unfortunately, however, mishaps with condoms are more likely to occur with anal-rectal intercourse than with vaginal intercourse. The use of a spermicidal jelly that contains nonoxynol-9 along with a condom is an added protection, because this spermicide also kills infected lymphocytes.

It is apparent that male-to-male and male-to-female transmissions are considered more likely than female-to-male transmission. Nevertheless, female-to-male transmission is also considered possible and may become more prevalent as more females become infected. One unhappy side effect to female infection is the fact that viruses and infected lymphocytes can pass to a fetus via the placenta or to an infant via mother's milk. Presently, infected infants account for about 1% of all AIDS cases.

SYMPTOMS OF AIDS

Most often when discussing the symptoms of AIDS, the discussion centers around the symptoms of full-blown AIDS. This is a mistake, because the full course of an HIV infection is not taken into account. It is important to realize that exposure to HIV and not membership in a risk group is the first step toward AIDS. Thereafter, it is possible to relate the progression of the disease to the number of T4 cells in the body, as is done in figure A.3 based on data from a particular individual. For the sake of our discussion, figure A.3 is divided into three stages:

Asymptomatic carrier—	first 9 months
Aids-related complex (ARC)—	10–63 months
Full-blown AIDS—	64–83 months

Asymptomatic Carrier

Usually individuals do not have any symptoms at all after initial infection. A few (2% to 10%) do have mononucleosis-like symptoms that may include fever, chills, aches, swollen lymph glands, and an itchy rash. These symptoms disappear, however, and there are no other symptoms for quite some time.

During the asymptomatic carrier stage the individual exhibits no symptoms and yet is highly infectious. According to figure A.4, first the blood contains a high amount of HIV, and then the body begins to produce antibodies against the virus. The standard HIV blood test indicates the presence of the antibody and not the presence of HIV itself. This means that an individual is infectious before the HIV test becomes positive.

A small percentage of false-negative and false-positive test results do occur with the HIV-antibody test. A positive result can be verified with a more expensive test that has a higher accuracy. Some people are not in favor

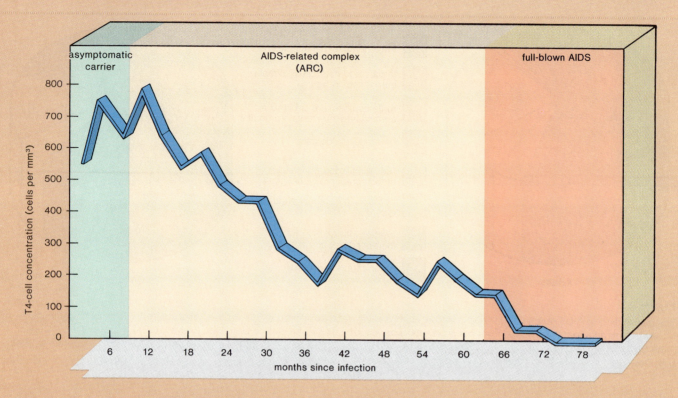

Figure A.3

T4-cell concentration in a young man whose HIV infection followed a typical course. After about 3 months the concentration of T4-cells increases as the body attempts to fight the infection. During this time, the individual is an asymptomatic carrier. Once the T4-cell concentration begins to decline, the individual develops the symptoms of ARC. Finally, full-blown AIDS is diagnosed when the patient has one or more of the opportunistic infections. By this time the T4-cell concentration is below 100 per cubic millimeter, and the individual soon dies. Source: Data from R. R. Redfield and D. S. Burke, "HIV Infection: The Clinical Picture," *Scientific American,* October 1988.

of routine testing for HIV antibodies because it could possibly subject those who test positive to unfair social discrimination. However, it is possible that routine testing could also help prevent the spread of AIDS.

It is important to realize that AIDS can be spread by persons who do not realize they are infected. Most young people who are infected contracted the virus from someone who did not know he or she was infected, and they themselves have not yet developed any symptoms and do not know they are infected.

AIDS-Related Complex (ARC)

Several months to a year after infection, the individual may begin to show some symptoms. The most common symptom of AIDS-related complex (ARC) is swollen lymph glands in the neck, armpits, or groin that persist for 3 months or more. Swollen lymph glands are believed to occur because B cells in the lymph glands are hyperactivated and produce a flood of antibodies in an effort to control the infection (fig. A.3).

Still, during this time, the number of T4 cells continues to decline. Symptoms that indicate that HIV infection is present are severe fatigue not related to exercise or drug use; unexplained persistent or recurrent fevers, often with night sweats; persistent cough not associated with smoking, a cold, or the flu; and persistent diarrhea. Also possible are signs of nervous system impairment, including loss of memory, inability to think clearly, loss of judgment, and/or depression.

When the individual develops non–life-threatening and recurrent infections, it is a signal that full-blown AIDS will occur shortly. One possible infection is thrush, a fungal infection that is identified by the presence of white spots and ulcers on the tongue and inside the mouth. The fungus may also spread to the vagina, resulting in a chronic infection. Another frequent infection is herpes simplex, with painful and persistent sores on the skin surrounding the anus, the genital area, and/or the mouth.

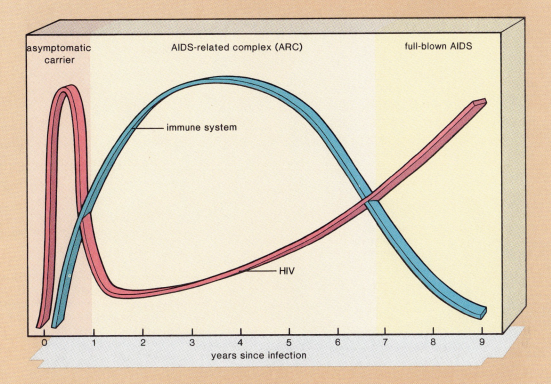

Figure A.4

Balance of power between HIV and the immune system during the course of an HIV infection. During the time that a person is an asymptomatic carrier the concentration of HIV in the body is high, but then the immune system becomes active. While an infected person has ARC the immune system begins to lose the battle, and eventually the battle is lost when a person develops full-blown AIDS. Source: Data from R. R. Redfield and D. S. Burke, "HIV Infection: The Clinical Picture," *Scientific American,* October 1988.

Full-Blown AIDS

Whereas previously it was believed that a percentage of persons with ARC would never progress to full-blown AIDS, it is now thought that almost every person with ARC will eventually develop full-blown AIDS. The number of persons with full-blown AIDS represents only the tip of the iceberg when the total number of persons who are infected with the HIV virus is considered (fig. A.5).

In this final stage of HIV infection, the AIDS patient, who is now suffering from "slim disease"—severe weight loss and weakness due to persistent diarrhea and coughing—will most likely succumb to one of the so-called opportunistic infections. These infections are called opportunistic because they are caused by microbes that cannot ordinarily start an infection, but in AIDS patients they have the opportunity because of the severely impaired immune system (fig. A.3). Some of the opportunistic infections are as follows:

Pneumocystis carinii **pneumonia.** The lungs become useless as they fill with fluid and debris due to an infection with this organism. There is not a single documented case of *P. carinii* pneumonia in a normal person; therefore, its presence establishes the diagnosis of AIDS in about 60% of patients.

Toxoplasmic encephalitis. This is caused by a one-cell parasite that lives in cats and other animals as well as humans. Many persons harbor a latent infection in the brain or muscle, but in AIDS patients the infection leads to loss of brain cells, seizures, weakness, or decreased sensation on one side of the body.

Mycobacterium avium. This infection affects many organs. Infection of the bone marrow can contribute to a decrease in red blood cells, white blood cells, and platelets in AIDS patients.

Kaposi's sarcoma. This unusual cancer of blood vessels gives rise to reddish purple, coin-size spots and lesions on the skin.

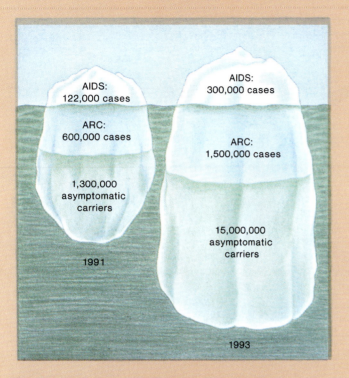

AIDS:
122,000 cases

ARC:
600,000 cases

1,300,000
asymptomatic
carriers

1991

AIDS:
300,000 cases

ARC:
1,500,000 cases

15,000,000
asymptomatic
carriers

1993

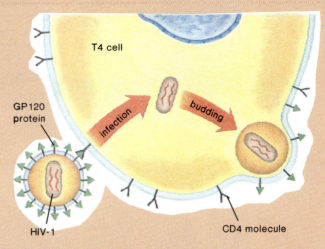

T4 cell

GP120
protein

infection

budding

HIV-1

CD4 molecule

Figure A.6
An HIV-1 has an envelope molecule called GP120, which allows it to attach to CD4 molecules that project from a T4 cell. Infection of the T4 cell follows, and HIV eventually buds from the infected T4 cell. If the immune system can be trained by the use of a vaccine to attack and destroy all cells that bear GP120, a person would not be infected with HIV-1.

Figure A.5
Most persons who have an HIV infection are asymptomatic carriers. They can pass the infection to others without even knowing that they are infected. A much smaller proportion of persons with an HIV infection have ARC and know that they are infected but are still able to carry on a fairly normal life. A still smaller proportion of those infected have been diagnosed as having AIDS. This diagnosis is based on the presence of an opportunistic disease. Sources: Data from Frank D. Cox, *The AIDS Booklet,* copyright © 1989 Wm. C. Brown Publishers, Dubuque, Iowa; and *HIV/AIDS Surveillance,* Year-End Edition, January 1990, U.S. Department of Health and Human Services.

Although drugs are being developed to deal with opportunistic diseases in AIDS patients, death usually follows in 2 to 4 years. Some infected individuals may continue to lead a fairly normal life for some months, but eventually they are hospitalized due to weight loss, constant fatigue, and multiple infections.

TREATMENT OF AIDS

Presently there is no cure for AIDS. Investigators have developed drug treatments that prolong the lives of those infected, and they are attempting to develop vaccines to prevent infection in the first place.

The drug zidovudine (AZT) has been shown to be effective in those with full-blown AIDS, and it also seems to prevent the progression of the disease in infected per-

sons who have fewer than 500 T4 cells per cubic millimeter but who exhibit no symptoms. A new drug called dideoxyinosine (DDI), which, like AZT, works by preventing viral replication in cells, is undergoing clinical trials and shows fewer side effects. Researchers are hopeful of having a protease inhibitor available soon. Protease enzymes are needed by HIV to reproduce. Also, new drugs are being developed to fight the opportunistic infections that AIDS patients contract. Controlling these infections could prolong the lives of AIDS patients.

Most AIDS cases in the United States are caused by HIV-1, and therefore the vaccines being developed are directed against this virus. Notice that the virus has an outer envelope molecule called GP120 (fig. A.6). When GP120 combines with a CD4 molecule that projects from helper T lymphocytes, it enters helper T cells. Two experimental vaccines for HIV-1 are being studied clinically. Both of these vaccines utilize only GP120, but one uses it directly and the other uses a treated vaccinia (cowpox) virus that has the GP120 gene inserted into it. The hope is that this gene will go on producing GP120 molecules after entering the host cell's blood. An entirely different approach is being taken by Jonas Salk, who discovered the polio vaccine. His vaccine utilizes whole HIV-1 killed by treatment with chemicals and radiation. So far, this vaccine has been found to be effective against the HIV-1 in chimpanzees.

AIDS PREVENTION

We have indicated that AIDS is transmitted by sexual contact or by sharing contaminated needles or blood. Shaking hands, hugging, social kissing, coughing or sneezing, and swimming in the same pool will not transmit the AIDS virus. You cannot get AIDS from inanimate objects like toilets, doorknobs, telephones, office machines, or household furniture.

The following behaviors will help prevent the spread of AIDS:

1. Do not use alcohol and drugs in a way that may prevent you from being able to control your behavior. Especially do not take up the habit of injecting drugs into veins.
2. If you are already a drug user and cannot stop your behavior, always use a clean, previously unused needle for injection.
3. Either abstain from sexual intercourse or develop a monogamous (always the same partner) sexual relationship with a partner who is free of HIV and is not an IV drug user.
4. If you do not know for certain that your partner has been free of HIV for the past 5 years, always use a latex condom during sexual intercourse. Be sure to follow the directions supplied by the manufacturer. Use of a spermicide containing nonoxynol-9 in addition to the condom can offer further protection because nonoxynol-9 also kills lymphocytes.
5. Refrain from multiple sex partners, especially with homosexual or bisexual men or IV drug users of either sex.

Aerial photo of Sun City, Arizona. Humans have the capability to drastically change the natural ecology of an area.

Evolution, Behavior, and Population Concerns

The way in which humans carry on sexual reproduction is an adaptation to life on land. It is possible to trace the evolution of sexual reproduction and in so doing compare the manner in which humans reproduce to the manner in which other organisms reproduce. The sexual behavior of humans can also be compared to that of other animals. It would be of interest to know if human sexual behavior is innate or largely learned.

Since 1850 the human population has expanded so rapidly that some doubt there will be sufficient energy and food to permit the same degree of growth in the future. Many would like to see the human population level off and recommend that energy be used efficiently and raw materials be recycled to ensure a continued supply. These measures will also help curtail pollution.

Evolution of Sexual Reproduction

Evolution, the process by which new types of organisms come into being, depends on the occurrence of genetic variations (genotype changes) that are of benefit to the species. Such variations enhance the likelihood of an organism's survival. **Survival of the fittest** does not mean that the "fit" personally destroy the "nonfit." Rather, the fit are more likely to survive and reproduce and pass on their genes to the next generation. What type of genetic variation is apt to make an organism more fit? A simple illustration of this is given in figure 12.1. Variations that cause the organism to be more adapted to the environment are those that increase the possibility of survival. Thus they are the ones that are more likely to be passed on. "Adapted" means that the organism's physical, physiological, and even behavioral characteristics are suited to the environment. The physical, physiological, and behavioral characteristics that enable a

Human fetus at 11–12 weeks.

fish to survive in water are those that make it suited or adapted to its environment. These are the features, or **adaptations,** that will allow a fish to survive and pass on its genes.

Evolution is a continuous natural process in which a certain genotype is passed on to the next generation to a greater extent than other genotypes. This genotype has been "selected" by the environment; hence, the process is called **natural selection.** In other words, organisms that possess the selected genotype are the ones that survive long enough to reproduce and pass on this genotype. Therefore, each successive generation will contain more organisms with the genotype.

While it seems almost ridiculous to say that water is wet and land is dry, this very fact is the dramatic environmental difference between life in the water and life on land. Life arose in the water because a water environment is less hazardous to living things. Cells are more than 50% water and are in constant danger of **desiccation,** or drying out, in a dry environment. Any organism living on the land must have a means of reproduction that prevents desiccation of the gametes and the zygote.

Figure 12.1

The evolutionary process explains how long-necked giraffes evolved.

Early giraffes probably had necks of various lengths.

Natural selection due to competition led to survival of the longer-necked giraffes and their offspring.

Eventually, only long-necked giraffes survived the competition.

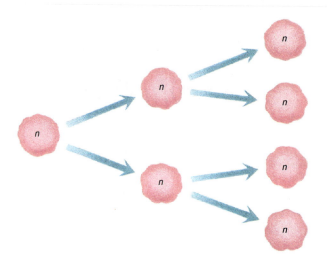

Figure 12.2

Not all animals practice sexual reproduction. In asexual reproduction, shown here for bacteria who are haploid (N), there is only one parent, and the offspring have the same genotype as the parent.

ASEXUAL REPRODUCTION

Sexual reproduction evolved, or came into being, during the course of the history of the earth. The first cell, believed to be a product of chemical evolution, simply divided to produce daughter cells. This type of **asexual reproduction** is common to bacteria today (fig. 12.2). Bacteria are haploid, and it may have been that the first cell was also haploid, meaning that it had only one set of genes. Such a life cycle seems inadequate for two reasons: (1) genetic variation is solely dependent on mutation, and (2) any faulty gene must express itself since the adult is haploid. For *bacteria* the system works well however, because numerous bacteria come into existence within a relatively short period of time. This means that actually the mutation rate is high, and a few defective genes will not reduce the total number of bacteria significantly.

SEXUAL REPRODUCTION

Origin of Sexual Reproduction

Sexual reproduction (reproduction by means of gametes) is first seen in primitive organisms that have a life cycle similar to that in figure 12.3. In the sexual portion of this life cycle, typically, the zygote "overwinters," meaning that it has a hard protective covering that allows survival during

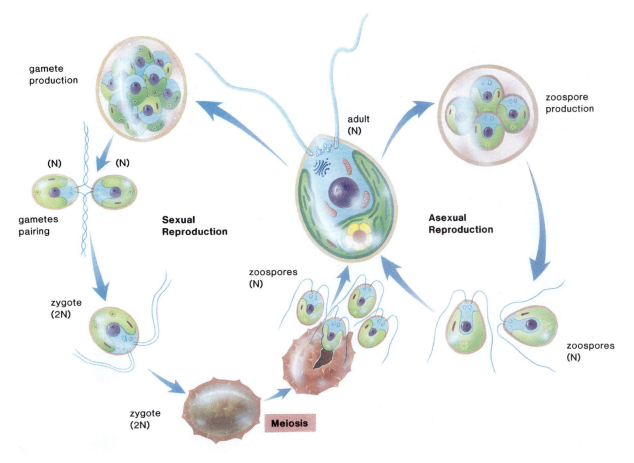

Figure 12.3

Life cycle of *Chlamydomonas*, an organism that undergoes both asexual reproduction and sexual reproduction. During the asexual cycle, the haploid adult divides to produce zoospores, each of which develops into an adult. During the sexual cycle, the haploid adult produces gametes that fuse, giving a zygote that develops a protective coat and overwinters. In the spring, meiosis produces haploid zoospores that develop into haploid adults.

the cold winter months. It may be theorized that this was an advantage to the species that favored the continuance of sexual reproduction. In sexual reproduction, genetic variation is routine because the offspring never has the same genes as either parent; instead, the offspring has a different combination of genes. Thus, sexual reproduction gives another method of achieving **genetic variation** in addition to chance mutation. Obviously, this increases the potential for possible adaptation, and sexual reproduction has become almost universal among living organisms. Table 12.1 contrasts asexual with sexual reproduction.

The life cycle depicted in figure 12.3, however, still has a haploid adult due to the fact that meiosis occurs after formation of the zygote. A diploid adult is preferred

TABLE 12.1   ASEXUAL VERSUS SEXUAL REPRODUCTION

	Asexual	Sexual
Number of parents	One	Two
Gametes	None	Gametes
Recombination of genes	Does not occur	Does occur

because with a double set of genes a defective gene may be masked or dominated by an effective gene. It is not surprising, then, that in the most recently evolved cycle the adult is diploid.

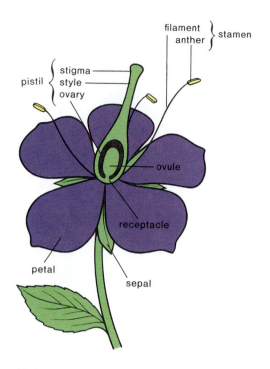

Figure 12.4
Anatomy of a flower. The pistil is the female part of the flower, and the stamens are the male parts.

Sexual Reproduction in Plants

Flowering plants have a life cycle that includes a diploid adult and sexual reproduction. Most higher plants have flowers in which there are male and female parts (fig. 12.4). The male parts of a flower, the stamens, have a portion called the anthers, which produce pollen. A pollen grain contains the male gamete, the sperm nucleus. The female part of the flower, called the pistil, includes an expanded portion, the ovary, that contains the ovules, each of which produces an egg. The manner in which the sperm nucleus reaches the egg nucleus in plants is an adaptation to life on land.

In plants, the hard outer covering of a pollen grain protects the sperm from drying out. In addition, a pollen grain is windblown or carried by insects to the top of the pistil. In other words, through the evolutionary process, reproduction in plants is not dependent on swimming sperm. This is advantageous because plants lack locomotion, and the male and female cannot get together for procreation as animals do.

Sexual Reproduction among Invertebrates

The life cycle of animals is shown in figure 1.3. Gametogenesis produces the gametes, the only haploid portion of the cycle. Since each parent contributes only half its genes to the zygote, the zygote has a different combination of genes than either parent. The adult is diploid and has a double set of genes.

Some animals are adapted to reproduction in the water, and some are adapted to reproduction on the land. This will be a major theme as we first survey a selection of invertebrates and then vertebrates.

Invertebrates are those animals that lack a backbone or vertebral column. We will look at some invertebrates that reproduce in the water and some that reproduce on land.

Flatworms

Freshwater *planarians,* or flatworms, can reproduce both asexually and sexually. During asexual reproduction the animal merely constricts below the pharynx and separates into two parts. Each part regrows the missing half.

Planarians are **hermaphroditic,** that is, they have both male and female organs (fig. 12.5). This is advantageous in that any two animals that meet can practice sexual reproduction, but it does not permit the specializations that can arise when the sexes are separate. The male organs are the testes (sperm production), the seminal vesicles (sperm storage), and a muscular penis. The female system consists of the ovaries and yolk glands, the oviducts, the vagina, and the seminal receptacle, which stores the sperm received.

When these animals copulate, each penis discharges sperm into the vagina of the other worm. After copulation the eggs with yolk are enclosed within a cocoon that is attached to a stationary object like a rock in the water. In marine forms the egg develops into a ciliated, free-swimming larva. A **larva** is an immature form that is capable of feeding and developing into the adult.

It might come as a surprise to learn that planarians have a penis, since this organ is a distinct advantage on land—it allows the direct passage of sperm from one animal to another and in that way protects the sperm from drying out. The use of a penis in aquatic animals means that they do not have to produce millions of eggs and sperm in order to increase the chances of fertilization.

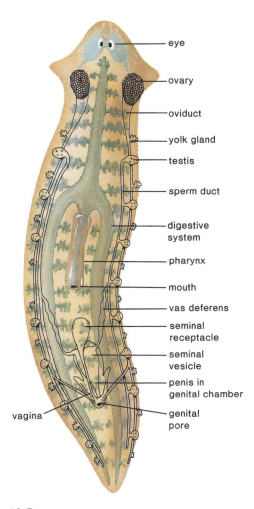

eye

ovary

oviduct

yolk gland

testis

sperm duct

digestive system

pharynx

mouth

vas deferens

seminal receptacle

seminal vesicle

penis in genital chamber

vagina

genital pore

Figure 12.5

Planarians are flatworms that live in fresh water. They are hermaphroditic and have both male and female sex organs. The female organs include an ovary, oviduct, yolk gland, and seminal receptacle. The male organs include a testis, sperm duct, vas deferens, seminal vesicle, and penis. The zygote develops into a swimming larva that can care for itself.

Tapeworms are flatworms that have taken up the parasitic way of life. They are intestinal parasites of humans that consist of a head region, a short neck, and then a long series of segments called proglottids. The head region contains only hooks and suckers for attachment to the intestinal wall. The proglottids each contain a full set of both male and female sex organs. Therefore, the tapeworm is little more than a reproductive factory in which self-fertilization is the rule. Obviously self-fertilization defeats the purpose of sexual reproduction—a means to increase the chance of variation among the offspring. However, the tapeworm's parasitic way of life makes it difficult to find another worm to reproduce with.

Figure 12.6

Earthworms are hermaphrodites that reproduce on land. During mating, the clitellum (smooth girdle) produces mucus that allows sperm to swim from one worm to the other. The mucus protects the sperm from drying out in the air.

Earthworms

Earthworms are also hermaphroditic, with a complete set of organs for both sexes. Mating occurs when two worms lie ventral surface to ventral surface, with the heads pointing in opposite directions (fig. 12.6). The clitellum, a smooth girdle about the worm's body, first secretes mucus, which holds the worms together and allows the sperm to swim from one worm to the other. Then the clitellum produces a slime tube that passes over the head of each worm. Eggs and sperm are deposited in the slime tube; fertilization results in zygotes, which develop directly into miniature earthworms. There is no larval stage.

Reproduction in the earthworm is not fully adapted to a land existence. There is no penis, and the sperm must make their own way from one worm to the other.

Grasshoppers

Reproduction in grasshoppers is typical of a land animal. The male has two testes and associated ducts that end in the penis. The female has ovaries that occupy the whole dorsal part of the animal, and there are accessory ducts that end in a vagina. The sperm received during copulation are stored in the seminal receptacles for future use. Fertilization is internal, usually occurring during late summer or early fall. The female deposits the fertilized eggs in the ground with the aid of her ovipositor.

TABLE 12.2	VERTEBRATES	
Type	**Example**	**Development**
Fish	Perch, trout, haddock	In water
Amphibian	Frog, salamander	In water
Reptile	Snake, turtle	Shelled egg on land
Bird	Chicken, penguin	Shelled egg on land
Mammal	Horse, elephant, ape, human	Within uterus of mother

In keeping with development on land, there is no larval stage. Instead there are a number of immature stages called nymphs. Nymphs are recognizable as grasshoppers even though they differ somewhat in shape and form from the adult.

Sexual Reproduction among Vertebrates

Vertebrates are those animals that have a backbone or vertebral column. The vertebrate animals are fishes, amphibians (e.g., frogs), birds, reptiles (e.g., snakes), and mammals (e.g., humans). Among vertebrates it is possible to see distinct differences between reproduction in the water and reproduction on land (table 12.2).

Fishes

Generally speaking, reproduction in the *fishes* requires external water. Sperm and eggs are usually shed into the water, where fertilization occurs. The zygote develops into a swimming larva that can fend for itself until it develops into the adult form.

Amphibians

The male *frog* lacks a penis and simply clasps the female during amplexus, causing her to release her eggs in the water. Thereafter, the male releases sperm (fig. 12.7).

Fertilization occurs externally in the water, and the zygote develops into a swimming larva (the tadpole) externally in the water. Metamorphosis produces the adult form that can move up onto the land.

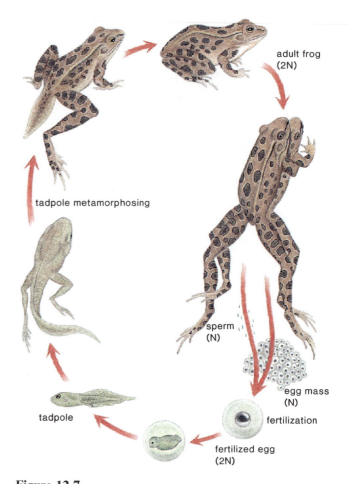

Figure 12.7

Life cycle of a frog. Frogs must return to water to reproduce because the gametes are shed in the water and the zygote develops into a swimming larval stage called the tadpole.

Reptiles

The *reptiles* were the first vertebrates to have a mode of reproduction suitable to the land environment. There is no need for external water to accomplish fertilization because the penis of the male passes sperm directly to the female. After internal fertilization has occurred, the egg is covered by a protective shell and laid in the ground. The shelled egg allowed development on land and eliminated the need for a swimming larva stage during development.

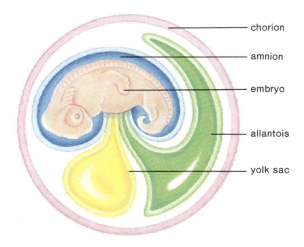

chorion
amnion
embryo
allantois
yolk sac

Figure 12.8
Reptiles have a shelled egg that allows reptiles to develop on land. The egg contains the extraembryonic membranes that protect and nourish the embryo so that a swimming larval stage is not necessary.

Figure 12.9
Newly born zebra still partially enveloped by afterbirth. When mammals develop they also have extraembryonic membranes (fig. 9.3), but these have been modified to allow internal development within the uterus.

• HUMAN ISSUE •

Much of what is known about early human development came from the study of invertebrates and other types of vertebrate animals. Is it beneficial, then, to accept that humans are related to other organisms through evolution, and that we can learn about humans by studying other animals? Or are there benefits to thinking of humans as not being related to other animals? What are the possible benefits?

For this to occur, the egg must have features that provide the developing embryo with oxygen, food, and water; that remove nitrogen wastes; and that protect it from drying out and from mechanical injury. This is accomplished by the presence of **extraembryonic membranes** (fig. 12.8). These membranes are not a part of the embryo itself and are disposed of after development is complete. There are four membranes: a yolk sac, which contains nourishing yolk; the allantois, which is a depository for nitrogenous waste; the chorion, which lies next to the shell and carries out gas exchange; and the fluid-filled amnion, which envelops the embryo and prevents it from drying out and protects it against mechanical injury. It may be noted that all vertebrates actually develop in the water—either external water as in the fish and amphibian or in amniotic fluid as in the reptiles, birds, and mammals.

Mammals

Most *mammals* are placental mammals that practice sexual reproduction in the same manner as humans. However, there are two other groups that are not placental mammals. The monotremes are egg-laying mammals represented only by the duck-billed platypus and the spiny anteater. The female incubates an egg and after hatching, the young are dependent upon the milk that seeps from glands on the abdomen. Another primitive group of mammals are the marsupials found in large numbers in Australia, such as the kangaroo and koala. The young are born prematurely, and they finish development in a pouch, where each attaches itself to a nipple.

In placental mammals the extraembryonic membranes of the reptilian egg have been modified for internal development within the uterus of the female. The chorion contributes to the fetal portion of the placenta, while a portion of the uterine wall contributes the maternal portion. Here nutrients, oxygen, and waste are exchanged between fetal and maternal blood. When these animals are born, the extraembryonic membranes are the afterbirth (fig. 12.9). These mammals not only have a long embryonic period, they are also dependent on their parents until the nervous system is fully developed and they have learned to care for themselves. ∎

SUMMARY

Evolution occurs when organisms better suited to the environment survive, reproduce, and pass on their genotype to the next generation. Since these organisms have been "selected by the environment," the process is called natural selection. Sexual reproduction among diploid organisms presumably evolved from asexual reproduction among haploid organisms. A diploid genotype increases the chances of survival and sexual reproduction increases the chances of genetic variation. Sexual reproduction has become adapted to the land environment in plants and animals.

KEY TERMS

adaptations 163

asexual reproduction 164

desiccation 163

evolution 163

extraembryonic membranes 169

genetic variation 165

hermaphroditic 166

invertebrates 166

larva 166

natural selection 163

sexual reproduction 164

survival of the fittest 163

vertebrates 168

REVIEW QUESTIONS

1. Define evolution, survival of the fittest, and adaptation.

2. What are two drawbacks to asexual reproduction?

3. How is genetic variation achieved during sexual reproduction?

4. How is reproduction in flowering plants adapted to a land environment?

5. Why could it be said that reproduction in the earthworm is not fully adapted to life on land?

6. How did evolution of the reptilian egg allow vertebrates to reproduce on land?

7. How does reproduction in placental mammals differ from reproduction in reptiles?

CRITICAL THINKING QUESTIONS

1. The tapeworm, a parasite in the gut of humans, is hermaphroditic, has no partner, and practices self-fertilization. Would you expect no variation among gametes? reduced variation? equal to that of having a partner?

2. How is reproduction in humans adapted to a land environment?

13

Biology of Sexual Behavior

The two basic aspects of behavioral patterns are diagrammed in figure 13.1. We assume that an animal's behavior is adaptive, that is, it increases the possibility that the genes will be passed on to the next generation. In other words, then, behavioral patterns are subject to evolution, just as other aspects of the phenotype are also subject to this process. The left-hand side of figure 13.1 concerns the evolution of behavior. The right-hand side of the figure concerns the execution of the behavioral pattern. The phenotype of the organism (a typical one for the species) includes a repertoire of behavioral patterns, and the particular stimulus determines which pattern will be executed. Often the organism must be ready to respond; sexual readiness or **motivation** is believed to be dependent, at least in part, on hormonal levels.

Animals with simple nervous systems tend to respond only to certain stimuli automatically in a predetermined way, whereas animals with complex nervous systems tend to learn to respond to selected stimuli with behavior suited to the particular circumstances. All animals, including humans, show some behavior of the first type, often called **innate** or instinctive behavior, but the higher animals have a larger component of learned behavior.

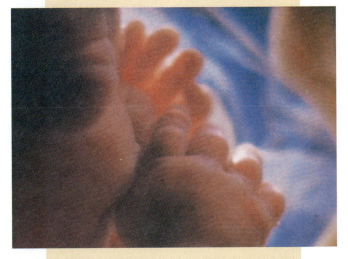

Close-up of human fetus sucking thumb at 3 months.

EXECUTION OF BEHAVIOR

Visual Stimuli

Sexual reproduction in lower animals is sometimes stimulated by visual stimuli. For example, male stickleback fish stake out a territory and build a nest; at this time their bodies become highly colored, including red bellies. Any male attempting to enter the territory is attacked as the owner repeatedly darts toward and nips the intruder. On the other hand, the owner entices a female to enter the territory by first darting toward her and then away in a so-called zigzag dance (fig. 13.2). Finally, he leads her to the nest, where she deposits her eggs. How does the male stickleback recognize another male as opposed to a female? Experimentation has shown that males attack any model that has a red underside, whether or not it resembles a fish. Therefore, the fish are believed to be programmed to respond automatically to sequential stimuli and are believed to carry out their reproductive behavior reflexively without the need for thought.

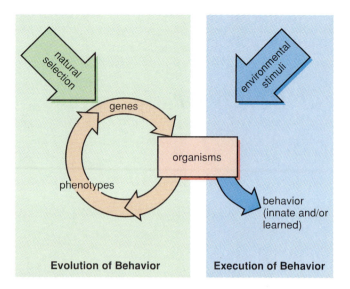

| Evolution of Behavior | Execution of Behavior |

Figure 13.1

Current interests in regard to behavior. *a.* Evolution of behavior. Selected phenotypes contribute genes to the next generation, and in this way behavior can evolve. *b.* Execution of behavior. Environmental stimuli cause organisms to display behavioral patterns that differ as to the degree they are innate or learned. In the latter case, the capacity to learn is a part of the phenotype.

It is possible that apes and humans also respond to visual more than other types of stimuli. Males are reported to be more sensitive to visual sexual stimuli than females, and it has been suggested that this is appropriate because an erection must occur before sexual intercourse is possible. However, these data are being challenged today since more women are willing to admit to being sexually aroused by visual and other stimuli.

Odor Stimuli

Lower animals are sexually stimulated by odors to a greater degree than humans. For example, female insects secrete chemicals called pheromones, and male insects are so attracted by the smell that they will travel great distances to find a solitary female. Since the male insects respond automatically, synthetic pheromones can be used to capture male insects, eliminating the necessity to spray many acres of land with insecticides. Owners of dogs are also aware that their pets are stimulated by odors.

Experiments are being conducted to determine if humans are at all sexually stimulated by odorous chemicals produced by the opposite sex. In one study researchers collected underarm secretions from men who wore a pad in each armpit. This "male essence" was then swabbed, three times a week, on the upper lips of seven

women whose uterine cycles typically lasted less than 26 days or more than 33. By the third month of such treatment the average length of the women's cycles began to approach the optimum 29.5 days—the cycle length associated with highest fertility. In another experiment, ten women with normal cycles were exposed to female underarm sweat in the same way as formerly. After 3 months the women's cycles were starting roughly in synchrony with those of the women who had donated the sweat. The researchers concluded that human beings are also subject to pheromone effects.

Sound Stimuli

Every year many kinds of frogs return to a specific mating pool. There the males begin to croak to signify that they are ready to mate. The North American carpenter frog sounds like a carpenter hammering nails, and a bullfrog's "h-h-rrumph" is familiar to most. A female of that species will approach the male if she is also ready to mate.

For humans, sounds such as music often set the appropriate mood for sexual excitement, and sounds uttered during a sexual experience contribute to that experience. Even so, sounds alone are not usually interpreted as sexually stimulating. The fantasizing that may accompany listening to music is classified as visual stimuli, since the subject is imagining different scenes in the mind's eye.

Touch Stimuli

Along with sight and sounds, touch represents a major source of sexual excitement. Specific areas of the body become erotic zones when these areas have been touched under circumstances judged to be sexual. In other words, we learn to have various erotic zones. Touch receptors accommodate to continuous stimuli so that they are no longer aroused; therefore, erogenous zones are better stimulated if they are stroked.

TERRITORIALITY AND DOMINANCE

Territoriality means that an animal defends a certain area, preventing certain other members of the same species from utilizing it (fig. 13.3). Territoriality spaces animals and therefore reduces aggression. It also avoids overcrowding, ensuring that the young will have enough to eat. Since some animals without a territory do not mate, it also has the effect of regulating population density to some extent. Males typically defend their territories against nonbreeding males, therefore territoriality may also help assure that only stronger males reproduce.

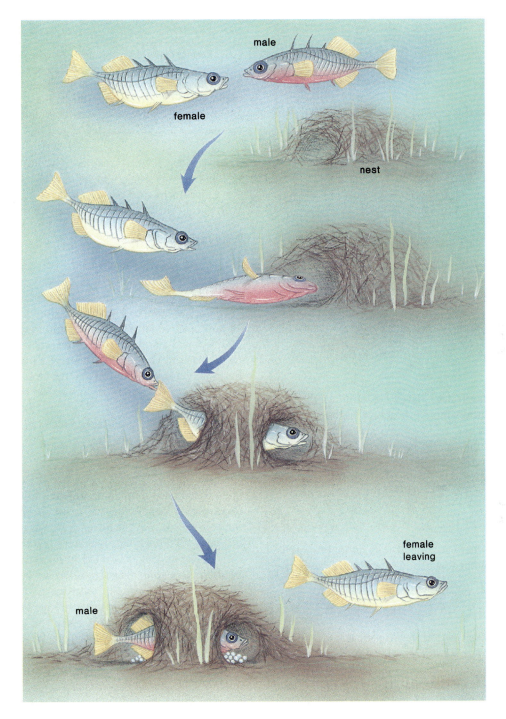

Figure 13.2

Mating behavior of stickleback fish. The male entices the female to the nest, where he lies flat on his side; the female swims into the nest and lays her eggs when prodded by the male, and then the male enters the nest to fertilize the eggs.

Figure 13.3
Two male elephant seals are battling over a piece of beachfront property. This will be the winner's territory, where he will collect a harem of females willing to mate with him. Territoriality partitions resources according to the supposed fitness of the animal.

A **dominance hierarchy** exists when animals within a society form a relationship in which a higher-ranking animal receives food and a chance to mate before a lower-ranking animal. Again, males have to defend their rank against the other males, and a dominance hierarchy may assure that only the stronger males reproduce.

MOTIVATION FOR SEXUAL BEHAVIOR

In lower animals it has been found that not only motivation but also subsequent sexual and reproductive activities are dependent on the presence of hormones.

Ring Doves

When male and female ring doves are separated, neither shows any tendency toward reproductive behavior. But when a pair are put together in a cage, the male begins courting by repeatedly bowing and cooing (fig. 13.4). Since castrated males do not do this, it can be reasoned that the hormone testosterone readies the male for this behavior. The sight of the male courting causes the pituitary gland in the female to release FSH and LH; these in turn cause her ovaries to produce eggs and release estrogen into the bloodstream. Now both male and female are ready to construct a nest, during which time copulation takes place.

•   HUMAN ISSUE   •

Are human beings like or different from other animals? Does the answer to this question have any bearing on whether or not human sexual behavior should be compared to that of other animals? What are the implications, if any, when it is possible to compare human sexual behavior to that of other animals?

The hormone progesterone is believed to cause the birds to incubate the eggs, and while they are incubating the eggs, the hormone prolactin causes crop growth so that both parents are capable of feeding their young crop milk.

Migrations

Hormones are also involved in the seasonal responses of animals. When the length of the day changes in the fall and spring, some birds (and other types of organisms) migrate long distances to their breeding grounds (fig. 13.5). In animals, we know that melatonin is produced at night by the pineal gland; therefore, the amount produced begins to increase in the fall and decrease in the spring. Melatonin is believed to inhibit the development of the reproductive organs, which accounts for why some animals reproduce in the spring. Reproduction in the spring is adaptive, because that is when food is available for growing offspring.

Humans

It has been known for some time that androgens (e.g., testosterone) promote the sex drive in both males and females. (Recall that the adrenal glands produce androgens in females.) This effect is believed to be due to the influence of the androgens on the brain rather than on the reproductive organs directly. This can be demonstrated in animals when a small hormone-containing pellet implanted in the hypothalamus of the brain increases sex drive. The amount of hormone in the pellet is too low to affect the reproductive organs directly; therefore, the effect must be directly on the brain. More recently it has been suggested that the secretion of the hormone oxytocin during love making contributes to the occurrence of orgasm and accounts for our sense of attachment to sex partners.

In humans, however, sequential reproductive activities may not be as dependent on hormonal levels because learning is so important to human behavior. In other words, in humans the presence or absence of hormones may cause motivation, but the actual activities are learned. The fact

Figure 13.4

Ring dove mating behavior. Successful reproductive behavior requires these steps: *a.* male performs courtship behavior as he bows and coos; *b.* suitable nesting materials and a site lead to building of the nest; *c.* copulation occurs; *d.* incubation of eggs; *e.* baby chicks are fed ''crop milk''; *f.* cycle begins again.

that humans learn reproductive behavior can be substantiated by the fact that different cultures have their own preferred reproductive and sexual behavior.

If human reproductive behavior is learned, then, isolated humans are expected to be awkward in performing sexual behavior. Experimentation of this type is not possible in humans but has been done in monkeys. Males raised in isolation are clumsy and ineffective when given a chance to mate. It would seem that although some human sexual and reproductive behavior is innate, much is also learned.

EVOLUTION OF BEHAVIOR

Is our sexual behavior ultimately controlled by the genes? Evidence that the genes do control behavior is provided by the nest-building habits of the parrot *Agapornis* (fig. 13.6). Three species of this genus have different methods of building nests. Females of the most primitive species use their sharp bills for cutting bits of bark, which they carry in their feathers to a site where they make a pad or a nest. Females of a more advanced species cut long regular strips of material, which are placed for transport only in the rump feathers. These strips are used to make elaborate nests with a special section for the eggs. Females of the most advanced species carry stronger materials, such as sticks, in their beaks. They construct roofed nests with two chambers and a passageway. The steady progression here from primitive to advanced suggests that an evolution of behavioral patterns has taken place.

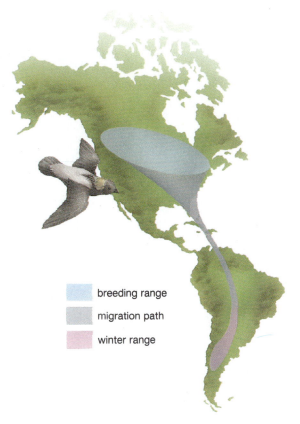

Figure 13.5
The bobolink is a bird that migrates from North America to Argentina in order to reproduce. The urge to migrate is believed to be due to hormonal changes that occur in the fall.

breeding range

migration path

winter range

Figure 13.6
Agapornis parrots transport nesting materials in different ways. Members of one species, shown here, cut strips of material and then tuck them into their rump feathers. Members of another species carry material in their beaks. When these two species are mated, the hybrids attempt to place material in the rump feathers and then in the end carry it in their beaks. This shows that behavior is controlled by genes.

The belief that genes control these traits can be supported by the following experiment. The species that carries material in its rump feathers is mated to the species that carries material in the beak. The resulting hybrid birds cut strips and try to tuck them in their rump feathers but are unsuccessful. It is as if the genes of one parent require them to tuck the strips, but the genes of the other species prevent them from succeeding.

An example that behavior is adapted for continuance of the genotype is found among the black-headed gulls, who always remove broken eggshells from their nests after the young have hatched. In a series of experiments, investigators determined that nests with broken shells remaining were subject to more predator attack than those from which shells had been removed. It can be reasoned that gulls that remove broken eggshells and other conspicuous objects from their nests will successfully raise more young than gulls that do not do so. This behavior pattern may have been selected through the evolutionary process because it improved reproductive success.

Sociobiologists believe that the sexual behavior of animals is ultimately controlled by the genes, and that sexual behavior is subject to evolution. The following examples demonstrate this reasoning.

Mammals

Mammals differ from other vertebrates in that the embryo develops within the uterus of the female. The behavior of mammals may very well be adapted to this situation, and the genes may dictate behavior that is consistent with this anatomical fact.

Female mammals produce many fewer eggs than males produce sperm. This difference in gamete production is not a disadvantage to females, because they have the assurance that the offspring is theirs—an assurance that males never have. Lacking physiological certainty, it behooves males of a species to attempt to impregnate as many females as possible. On the other hand, since pregnant females need the protection of males, it seems consistent that they would be faithful. Does this description also pertain to humans? Is this why some people believe that males have more sexual freedom than females?

Female mammals usually care for the young they bear because they know the offspring definitely carry one-half of their genes (fig. 13.7). Males may not contribute as greatly to the care of the young because they have no physiological evidence that the offspring belongs to them

Figure 13.7

Among baboons, females and not males care for the young. Since females give birth to offspring, they are certain that an offspring is theirs.

and that it carries one-half of their genes. Since females need protection while they care for their offspring, they tend to select aggressive males on which to be dependent. The evolutionary outcome is that females are docile and subservient, while males are combatant and dominant. Again, can we find correlations in human behavior?

Humans

According to sociobiologists who have extended the theory of the evolution of sexual behavior to humans, males have traditionally been valued for their position in the world and females for their ability to reproduce. Also traditionally, young women are apt to marry older males because they are capable of providing for a family, while the young woman is capable of producing offspring.

Figure 13.8

Human family group. Investigators are trying to determine how much of our sex role behavior is due to genes.

Humans are the only mammal lacking an estrus, commonly known as the period of being in heat. The female is continuously amenable to sexual intercourse. Could this possibly be an adaptation to the necessities of reproduction? Human infants are helpless and require the committed attention of at least one parent, usually the woman. The mother and child need the protection of the father for an extended length of time (fig. 13.8). Does the fact that sex is continuously available help assure a continued relationship between male and female as they both contribute to the upbringing of the offspring?

Males tend to be aggressive, but keep this aggression under control, particularly with persons they recognize as relatives. Knowledge of kinship is very important to humans. Altruism, or self-sacrificing, exists but more so for a relative than a nonrelative. In other words, the degree to which others share our genes influences our behavior toward them.

Interestingly enough, various marriage habits around the world suggest a possible adaptation to the environment. Among African tribes, one man may have several wives. This is reproductively advantageous to the male, but it is also advantageous to the woman, since by this arrangement she has fewer children, who are thereby assured a more nutritious diet. In Africa, sources of protein are scarce, and protein deficiency diseases are a threat to continued existence.

Among the Bihari hillmen of India, brothers have one wife. The environment is hostile, and it takes two men to provide the necessities for one family. The fact that the men are brothers means that they share genes in common, and therefore they are actually helping each other look after common genes.

Experiments in Human Reproductive Behavior

Experiments in human social behavior are few, but the kibbutz in Israel is one such example. Here, a real attempt has been made to do away with the usual sex role differences. Men and women work alongside one another, and children are raised together in separate living quarters. Even so, there has developed a division of labor according to sex. The men do the more strenuous laboring jobs, and the women do the customary "women's work"—washing clothes and preparing meals. Men tend to seek out positions of authority, whereas the women do not. Also, the women take more interest in their children, visiting them daily to establish a relationship that would not otherwise develop.

SUMMARY

Sexual behavior occurs as a response to environmental stimuli. This is best demonstrated in lower animals but most likely also applies to humans. Hormones are believed to affect motivation for sexual behavior. The exact response humans make, however, is largely learned.

Evidence suggests that animal behavior is inherited and subject to the evolutionary process. Behavior that increases the chances of having offspring is favored. The way in which mammals reproduce seems to influence male/female sex roles.

Examples also show that human sexual behavior is suited to the environment. Thus, the possibility exists that human sexual behavior (including sexual roles) may be adaptive, making it difficult for humans to assume roles other than the traditional ones. Sociobiologists believe that human male and female behavior has evolved to increase the chances of each sex passing their genes on to the next generation.

KEY TERMS

dominance hierarchy 174

innate 171

motivation 171

sociobiologists 177

territoriality 172

REVIEW QUESTIONS

1. Explain the diagram in figure 13.1.

2. Give examples of external stimuli that promote sexual reproductive behavior in animals and humans.

3. What is territoriality and a dominance hierarchy? What role do they play in reproductive strategies?

4. Give evidence that the reproductive behavior of ring doves is controlled by hormones.

5. Discuss the control of sexual behavior in humans. Are hormones involved? innate responses? learned responses?

6. Describe an experiment that shows behavior is inherited.

7. What does sociobiology have to say about human reproductive behavior? Is there evidence to support these ideas?

CRITICAL THINKING QUESTIONS

1. What evidence exists and what more evidence might you need to support the belief that human sexual behavior is dependent on hormone levels in the body?

2. For what reasons might feminists be opposed to the suggestion that male and female sex roles are inherited rather than learned?

Population Concerns

The human population ever increases in size, and this growth places a burden on the environment. The environment supplies us with the resources we need and also takes up pollutants. If the environment is stressed, the human population itself is threatened.

HUMAN POPULATION GROWTH

The human growth curve is an exponential curve (fig. 14.1). In the beginning, growth of the population was relatively slow, but as greater numbers of reproducing individuals were added to the population, growth increased until the curve began to slope steeply upward. The slope of the curve at any one point shows how fast the population was increasing in size at that time. It is apparent from the position of 1990 on a growth curve of the human population that growth is now quite rapid.

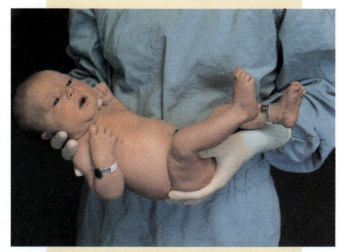

A newborn, 7 lb. 14 oz.

At current rates, the world adds the equivalent of a medium-sized city every day (200,000) and the combined populations of the United Kingdom, Norway, Ireland, Iceland, Finland and Denmark every year.

Growth Rate

The **growth rate** of a population is determined by subtracting the number of people who die from the number of individuals who are born per 100 persons per year. For example, if the death rate is 1.0 individuals while the birthrate is 3.0 individuals per 100 persons per year, then the growth rate for the year would be 2.0 individuals per 100 persons, or 2%:

$$G.R. = \frac{3.0 - 1.0}{100}$$
$$= 2\%$$

The growth rate for the world, which reached almost 2%, has declined slightly, and yet the world population still increases in size because you must multiply the growth rate by the present population size in order to calculate the increase. For example, compare the annual increase of population A, composed of 100 persons with a growth rate of 1.5%, to population B, composed of 2,000 persons with a growth rate of 1.3%.

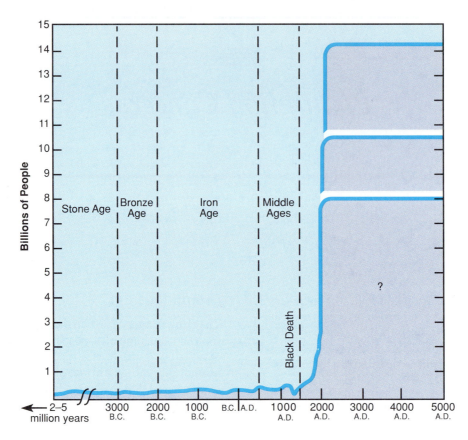

Figure 14.1

The human population has been undergoing rapid exponential growth. It is predicted that the population size will level off at 8, 10.5, or 14.2 billion, depending upon the growth rate.

Population A
Population size = 100 persons
Growth rate = 1.5%
Annual increase = 1.5 persons

Population B
Population size = 2,000 persons
Growth rate = 1.3%
Annual increase = 26 persons

In other words, a small population with a large growth rate will not increase dramatically, whereas a large population with a modest growth rate is capable of increasing dramatically. The present world population is large—in 1650 there were about 500 million people in the world; in 1830 there were 1 billion; in 1930 there were 2 billion; in 1960 there were 3 billion; and by 1990 there were over 5 billion. Since the world population is large, even a modest growth rate adds 85 million individuals to the population each year (fig. 14.2).

With a growth rate of 1.8%, the world population is expected to double within 39 years. The **doubling time** for a population is calculated by dividing 70 (the demographic constant) by the growth rate:

$$\text{D.T.} = \frac{70}{1.7} \simeq 40 \text{ years}$$

Many persons question if the world will be able to support the 10 billion persons expected by the year 2030 (1990 + 40) if this doubling occurs. In other words, they ask, What is the carrying capacity of the earth?

Carrying Capacity

If the growth curve for nonhuman populations is examined, the population often tends to level off at a certain size. For example, figure 14.3 gives the actual data for the growth of a fruit fly population reared in a culture bottle. At the beginning, the fruit flies were adjusting to their new environment and growth was slow. But then, since food and space were plentiful, they began to multiply rapidly. Notice that the curve begins to rise dramatically just as

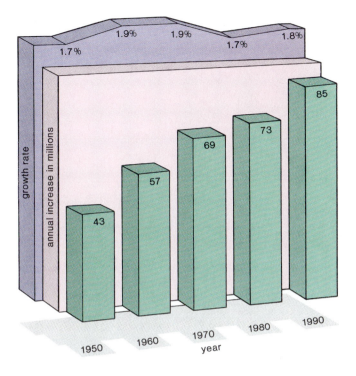

Figure 14.2

World population growth by decade. Since 1950 the growth rate has varied between 1.7% and 1.9%. Each decade has had more of an increase than the previous decade despite a decline in the growth rate from a high of 1.9%. Source: Data from the Population Reference Bureau, Inc.

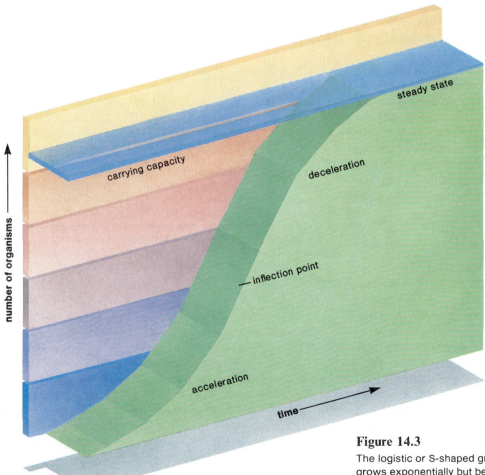

Figure 14.3

The logistic or S-shaped growth curve. A population initially grows exponentially but begins to slow down as resources become limited. Population size remains stable when the carrying capacity of the environment is reached.

the human population curve does. At this time, it may be said that the population is demonstrating its **biotic potential,** which is the maximum growth rate under ideal conditions. Biotic potential is not usually demonstrated for long because of an opposing force called **environmental resistance.** Environmental resistance includes all the factors that cause early death of organisms and thus prevent the population from producing as many offspring as it might otherwise have done. As far as the fruit flies are concerned, we can speculate that environmental resistance included the limiting factors of food and space. Also, the waste given off by the fruit flies may have contributed to keeping the population down.

The eventual size of any population represents a compromise between the biotic potential and environmental resistance. This compromise occurs at the **carrying capacity** of the environment. The carrying capacity is the maximum population that the environment can support.

Experts have a difference of opinion as to the carrying capacity of the earth in regard to humans. Some authorities think the earth is capable of supporting 50 to 100 billion people. Others think we already have more humans than the earth can support.

More Developed and Less Developed Countries

While the growth rate and doubling time for the world is 1.8% and about 40 years, respectively, each individual country can be studied separately. Countries that tend to be less industrial are classified as **less developed countries,** and countries that are industrial are classified as **more developed countries.** The less developed countries contribute the most to the world's growth rate. When the growth rate is analyzed for these countries, it is found that the birthrate is proportionately higher than the death rate. This causes the growth rate in these countries to increase and the population size to increase dramatically. It is estimated that in the year 2000 the population of the more developed countries will be about 1.3 billion, while the

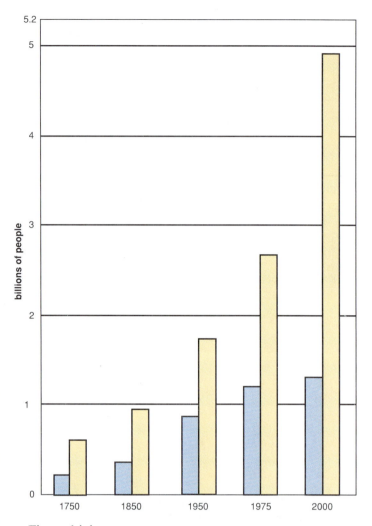

Figure 14.4

Size of human population in more developed versus less developed countries. The population of the developed countries (blue) increased between 1850 and 1950, but the size is expected to increase little between 1975 and 2000. In contrast, the population size of the less developed countries (yellow) increased in the past and is expected to increase dramatically in the future also. From "The Populations of the Underdeveloped Countries" by Paul Demeny. Copyright © 1974 by Scientific American, Inc. All Rights Reserved.

population of the less developed countries will be about 4.9 billion (fig. 14.4).

At current growth rates, most less developed countries will double their population within 25 years, and therefore they will have to double their economic output within 25 years to keep their present standard of living. It is estimated that India needs 4 million new jobs each year in order to maintain the present standard of living, without even considering improving the standard of living.

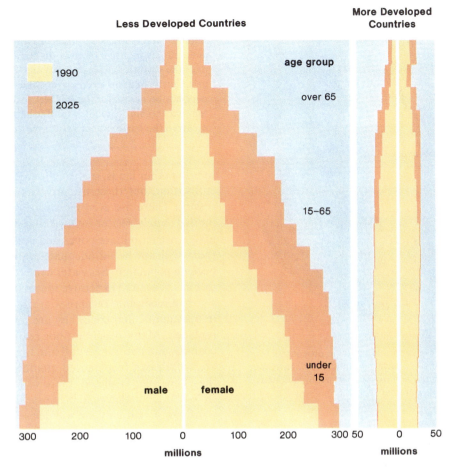

Figure 14.5
Age structure for less developed and more developed countries. The less developed countries contain 75% of the world population versus 25% in the more developed countries. The less developed countries will continue to expand for many years because of their youthful profile, but the more developed countries are approaching stabilization. From "The Growing Population" by Nathan Keyfitz. Copyright © 1989 by Scientific American, Inc. All rights reserved.

Zero Population Growth

In every country, people live longer than they used to; modern medicine and technology have increased the life span of all. Almost everyone would agree that the desirable way to bring about **zero population growth** is not to suddenly allow older people to die but to decrease the birthrate instead.

It is commonly believed that if each couple replaced only themselves (one couple having two children), then the population size would be stabilized at its present level. This is not the case, however, because of the age composition of most populations.

Figure 14.5 is an age structure diagram for a less developed country and for a more developed country. The diagram for a less developed country has a pyramid shape typical of a young and expanding population in which there are more young women entering the reproductive

years than older women leaving those years behind. The rectangular-shaped **age structure diagram** for a more developed country, on the other hand, is that of a stable population inasmuch as there are approximately the same number of persons in all three age categories. The United States falls somewhere between these two extremes and does have more women entering the reproductive years than those leaving them. This means that the United States will continue to have a population increase even if reproduction replacement were put into practice.

RESULTS OF POPULATION INCREASE

Increased population has two general effects: (1) it increases the consumption of natural resources, and (2) it causes an increase in pollution.

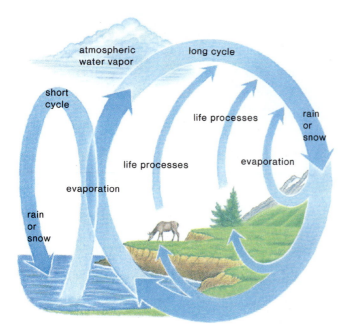

Figure 14.6

Short and long water cycles. Evaporation and life processes return water to the atmosphere, which then falls to the earth in the form of rain or snow. Water is a renewable resource because a new supply is always being made available in this way.

Resource Consumption

Natural resources are classified as renewable and nonrenewable. Examples of **renewable resources** are food and water. Every time crops are planted, more food is expected. Every time it rains, there is more water; the water we use up is returned to us because of water cycles (fig. 14.6). The definition of renewable resources might cause us to think that there is no reason to worry about the supply of these resources. But we are now very much aware of the fact that the quantity of renewable resources cannot be expanded forever. Even now we do not produce and equitably distribute enough food to feed all persons adequately. And the possibility of producing and distributing enough to feed 10 billion seems doubtful.

All the lands that we now know are capable of cultivation are under cultivation. Attempts have been made

to bring other lands, such as the tropics and deserts, under cultivation, but these attempts have as yet been unsuccessful. There are even indications that fresh water is in limited supply. Water is now being diverted from less populated areas to more populated areas. This suggests that there will not be an adequate water supply if all areas become populated. Seawater can be desalted, but thus far this has proven so expensive that it is not done extensively.

Nonrenewable resources are those that exist in limited amounts, and once they are used up they cannot be replaced. Many people do not realize that fuels such as coal, oil, and natural gas fall into this category. These fuels are withdrawn from stored quantities in the earth; once they have been depleted, there will be no more. It is possible that technology can help ease the problem of possible shortages. Eventually, we will devise means to make usable energy from renewable resources. For example, oil and coallike pellets can be made from garbage, but it may take time to perfect the process. Also with time, solar energy can replace fossil fuels for certain purposes, such as heating our homes. At present little is being done to develop these alternative ways of acquiring energy.

Technological methods can increase the amount of food available by developing new breeds of plants that produce a higher yield than the parent plants. But to this day, these plants require increased amounts of fertilizer and frequent spraying with pesticides. Both of these procedures eventually cause pollution.

Pollution

Technology goes hand in hand with pollution. The largest polluter of air is not factories but the automobile (figure 14.8). The more people, the more cars on the road and the

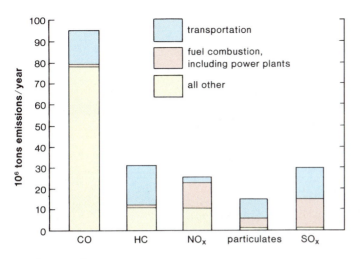

Figure 14.7

Components of air pollution: CO (carbon monoxide), HC (hydrocarbons), NO_x (nitrogen oxides), particulates (solid matter), SO_x (sulfur oxides). Transportation contributes most to air pollution because when gasoline burns it gives off CO, a gas that interferes with the capacity of hemoglobin to carry oxygen.

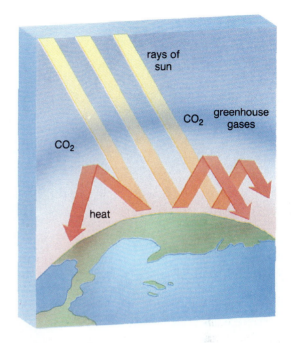

Figure 14.8

The greenhouse effect. The sun's rays can penetrate CO_2 and reach the earth's surface, but the resulting heat cannot pass through a layer of CO_2 and instead is trapped near the earth's surface, just as heat is trapped by the glass of a greenhouse. The earth's temperature is forecast to increase substantially due to the accumulation of CO_2 in the atmosphere.

greater the pollution. Even with stricter emission standards today, the amount of pollution from automobiles stays about the same because of the increased numbers of automobiles. Air pollution gases not only from automobiles but also from power plants cause acid rain, which is destroying trees in forests and living things in our lakes. These gases also contribute to a gradual warming of the earth, known as the greenhouse effect (fig. 14.9). The greenhouse effect is of concern, because if the earth warms the sea will rise and most major cities of the world will be flooded; also, the best temperature for growing crops will be farther north, where the soil is not as fertile.

Humans, as biological creatures, produce pollution. Many of our rivers and harbors today are open sewers, carrying not only waste from factories but also human waste. It is against the law in most areas to dump raw sewage into rivers, but the process continues in areas where no laws exist or where enforcement is lacking. Every city and town needs a sewage treatment plant. In a primary treatment plant, solids, grease, and scum are removed; in a secondary treatment plant, any remaining organic matter

is dissolved by bacterial action; in a **tertiary sewage treatment plant,** nutrient molecules are removed to prevent overgrowth of the water's natural inhabitants, such as algae. Tertiary treatment plants are very expensive, but it is possible to use alternative methods of removing nutrient molecules. The water leaving a secondary treatment plant can be passed through ponds, where aquatic plants will remove these nutrient molecules, for example.

Another type of pollution of concern is the loss of open spaces not only for ourselves but for the plants and animals that share this planet with us. Citizens of countries in the Southern Hemisphere are destroying the tropical rain forests because they need land on which to grow crops, or they can turn the trees into charcoal to fire the furnaces

Figure 14.9

Environmental preservation is an important consideration because natural areas perform many services for us. For example, they help keep pollution under control and provide a place for much needed human recreation.

of nearby industrial plants. The end result will be extinction of many plants and animals. The preservation of natural areas should be of concern to all humans for the following reasons:

1. Organisms now becoming extinct may be useful to human beings in the future. While emphasis is often placed on saving vertebrates from extinction, plants and invertebrates are just as necessary to human well-being.

2. Human beings seem to have an innate need to visit, at least occasionally, a diversified natural area. As Jon Roush has stated, "When people talk about going to the country for a vacation, they do not mean simply getting out of town. They mean finding some version of diversity. A cornfield may be natural, but it is also monotonous, and hardly anyone vacations there."[1]

3. Future generations have a right to inherit a world as complete as possible. In the same vein, other living things have a right to live in their natural environment rather than as captives in zoos. To be a polar bear, the animal must live like a polar bear; to be a chimpanzee, the animal must develop and mature as its ancestors in the wild have done.

4. Finally, we must realize that humans depend on natural areas (fig. 14.9). Natural areas not only have the capacity to absorb pollutants if not overwhelmed by them, they also contain the complexity that is necessary to stabilize the environment. It is in our own best interests to retain and if necessary to restore what is left of natural areas. ∎

1. G.J. Roush, "On Saving Diversity," *The Nature Conservancy News* (January–February 1982), p. 8.

SUMMARY

At the present time, the population of the world is undergoing exponential growth; therefore, even though the annual growth rate has declined slightly, the world population still continues to increase dramatically. Most of this increase will occur in the less developed countries rather than the more developed ones. The less developed countries tend to have a pyramid-shaped age profile, indicating that a very large number of individuals are young and will be entering the reproductive years. Therefore, replacement reproduction would not immediately bring about zero population growth.

Whether or not the earth is already overpopulated is a serious question. We are incapable of feeding the masses; water, fossil fuels, and other resources are also in short supply. While many believe that technology can overcome all these difficulties, we now know that technology itself can create ecological problems. Since the earth has a finite carrying capacity, it might be well to realize that the human population cannot continue to increase indefinitely.

KEY TERMS

age structure diagram 185

biotic potential 184

carrying capacity 184

doubling time 182

environmental resistance 184

growth rate 181

less developed countries 184

more developed countries 184

nonrenewable resources 186

renewable resources 186

tertiary sewage treatment plant 187

zero population growth 185

REVIEW QUESTIONS

1. What is the formula for growth rate? If the birthrate is 0.9% and the death rate is 0.2% for a certain population, what would be the growth rate?

2. What is the formula for doubling time? What is the doubling time for the population described in question 1?

3. Explain why a small population rather than a large population can tolerate a high growth rate and a short doubling time.

4. Draw a growth curve that includes biotic potential, environmental resistance, and carrying capacity.

5. How do less developed countries differ from more developed countries in terms of industrialization, growth rate, and doubling time?

6. What is zero population growth? Why can't it be achieved now in the United States by each couple having two children?

7. What are the two general effects of increased population? Give examples of both.

8. What are two major concerns today in regard to air pollution?

9. Why should we preserve natural areas?

CRITICAL THINKING QUESTIONS

1. Can you think of reasons why it might be inappropriate to suggest that the growth curve in figure 14.3 also pertains to the human population?

2. The Mexican government at one time encouraged large families because it believed that the greater the number of people, the greater the work force, and the greater the prosperity. What is wrong with this thinking?

FURTHER READINGS FOR PART III

Alcock, J. 1984. *Animal behavior: An evolutionary approach,* 3d ed. Sunderland, Mass.: Sinauer.

Ayala, F. J. September 1978. The mechanisms of evolution. *Scientific American.*

Barash, D. P. 1982. *Sociobiology and behavior,* 2d ed. New York: Elsevier North-Holland.

Batie, S. S., and R. G. Healy. February 1983. The future of American agriculture. *Scientific American.*

Brown, L. R., et al. 1991. *State of the world: 1991.* New York: W. W. Norton and Co.

Bushbacher, R. J. January 1986. Tropical deforestation and pasture development. *BioScience.*

Ecology, evolution and population biology. 1974. Readings from *Scientific American.* San Francisco: W. H. Freeman & Co.

Ehrlich P., and A. Ehrlich. 1981. *Extinction.* New York: Random House.

Energy for planet Earth. September 1990. *Scientific American,* Special Issue.

Futuyma, E. J. 1986. *Evolutionary biology,* 2d ed. Sunderland, Mass.: Sinauer.

Gwatkin, D. R. May 1982. Life expectancy and population growth in the third world. *Scientific American.*

Houghton, R. A., and G. M. Woodwell. April 1989. Global climatic change. *Scientific American.*

Jones, Phillip D. and Tom M. L. Wigley. August 1990, Global warming trends. *Scientific American.*

Lewontin, R. C. September 1978. Adaptation. *Scientific American.*

Managing planet Earth. September 1989. *Scientific American,* Special Issue.

Miller, G. T., Jr. 1988. *Living in the environment: An introduction to environmental science,* 5th ed. Belmont, Calif.: Wadsworth.

Mohnen, V. A. August 1988. The challenge of acid rain. *Scientific American.*

Nebel, B. J. 1987. *Environmental science: The way the world works,* 2d ed. Englewood Cliffs, N.J.: Prentice Hall.

O'Leary, P. R., P. W. Walsh, and R. H. Ham. December 1988. Managing solid waste. *Scientific American.*

Power, J. F., and R. F. Follett. March 1987. Monoculture. *Scientific American.*

Revelle, R. August 1982. Carbon dioxide and world climate. *Scientific American.*

Savage, J. M. 1977. *Evolution,* 3d ed. New York: Holt, Rinehart, & Winston.

Volpe, P. E. 1985. *Understanding evolution,* 5th ed. Dubuque, Iowa: Wm. C. Brown.

White, R. M. July 1990. The great climate debate. *Scientific American.*

Acronyms

A, adenine (p. 143)

ACTH, andrenocorticotropic hormone (p. 74)

ADH, antidiuretic hormone (p. 74)

AID, artificial insemination by donor (p. 140)

AIDS, acquired immune deficiency syndrome (p. 144)

ARC, AIDS-related complex (p. 144)

bGH, bovine growth hormone (p. 58)

BLG, betalactoglobulin (p. 60)

C, cytosine (p. 43)

DDI, dideoxyinosine (p. 158)

DMD, Duchenne muscular dystrophy (p. 39)

DNA, deoxyribonucleic acid (p. 43)

FSH, follicle-stimulating hormone (p. 90)

G, guanine (p. 43)

GH, growth hormone (p. 181)

GIFT, gamete intrafallopian transfer (p. 141)

GnRH, gonadotrophic-releasing hormone (p. 78)

HCG, human chorionic gonadotrophin (p. 112)

HD, Huntington disease (p. 26)

HDN, hemolytic disease of the newborn (p. 31)

HGH, human growth hormone (p. 60)

HIV, human immunodeficiency virus (p. 144)

HMG, human menopausal gonadotrophin (p. 102)

HPV, human papilloma virus (p. 146)

HSV, *Herpes Simplex* virus (p. 145)

ICSH, interstitial cell stimulating hormone (p. 90)

IUD, intrauterine device (p. 133)

IUI, intrauterine insemination (p. 140)

IV, intravenous (p. 153)

IVF, in vitro fertilization (p. 141)

LH, luteinizing hormone (p. 103)

mRNA, messenger RNA (p. 46)

NF, neurofibromatosis (p. 26)

NGU, nongonococcal urethritis (p. 148)

PCR, polymerase chain reaction (p. 61)

PID, pelvic inflammatory disease (p. 134)

PKU, phenylketonuria (p. 26)

RFLPs, restriction fragment length polymorphism (p. 63)

RNA, ribonucleic acid (p. 46)

rRNA, ribosomal RNA (p. 46)

SCID, severe combined immune deficiency (p. 39)

STD, sexually transmitted disease (p. 143)

T, thymine (p. 43)

TDF, testis-determining factor (p. 35)

tPA, tissue plasminogen activator (p. 58)

tRNA, transfer RNA (p. 46)

TSH, Thyrotropic-stimulating hormone (p. 76)

U, uracil (p. 45)

Glossary

 A

abortion premature expulsion of an embryo or fetus from the uterus 137

acquired immunodeficiency syndrome (AIDS) a sexually transmitted disease caused by HIV that reduces the number of helper T-cells in the blood, impairing the immunity so that the patient succumbs to opportunistic diseases 144

acrosome a cap over the sperm head that contains enzymes believed to assist the process of fertilization 90

adaptation changes by which an organism becomes fit for its environment in order that it may survive and reproduce 163

adenine (A) one of four nucleotide bases found in DNA and RNA; pairs complementarily with thymine (T) in DNA and uracil (U) in RNA 43

adrenal gland endocrine gland that lies atop the kidneys 74

adrenocorticotropic hormone (ACTH) hormone secreted by the anterior lobe of the pituitary gland that stimulates activity of adrenal cortex 74

age structure diagram a diagram based on prereproductive, reproductive, and postreproductive age categories that tells whether a population will be stable, will expand, or will decline in size 185

AIDS-related complex (ARC) a set of symptoms, including swollen lymph nodes and night sweats, that precede the development of full-blown AIDS 144

allantois one of the extraembryonic membranes that in humans is a source of umbilical blood vessels to and from the placenta 114

allele an alternative form of a gene that occurs at a given chromosome site (locus) 21

amino acid a subunit of a protein 48

amniocentesis the removal of a small amount of amniotic fluid to examine the chromosomes and the enzymatic potential of fetal cells 117

amnion the extraembryonic membrane containing the amnionic fluid that bathes the embryo and fetus 113

androgens male sex hormones of which testosterone is the most potent 78

antibiotic a medicine that specifically interferes with bacterial metabolism and in that way cures humans of a bacterial disease 143

antibody chemical made by an individual to react with and thereby combat antigens 143

anticodon a "triplet" of three nucleotide bases in transfer RNA that pairs with a complementary triplet (codon) in messenger RNA 50

antidiuretic hormone (ADH) a hormone secreted by the posterior pituitary that controls the rate at which water is reabsorbed by the kidneys 74

areola a colored ring of tissue surrounding the nipple of the breast 83

artificial insemination the placement of sperm by instrumentation in the female vagina or uterus for the purpose of causing fertilization and pregnancy 140

asexual reproduction reproduction that requires only one parent and does not involve gametes 164

autosome any chromosome other than sex chromosomes. In humans there are 22 pairs of autosomes 5

 B

bioengineered the insertion of a foreign gene into DNA or an organism where it functions normally 55

biotechnology use of a natural biological system to make a new product or achieve a desired end 55

biotic potential the reproductive ability of a population when environmental conditions are ideal 184

birth-control pill medication in pill form that contains estrogen and progesterone for the purpose of preventing pregnancy 130

C

cancer irregular and uncontrolled growth of cells 53

Candida albicans the causative agent for yeast infection of the vagina; common in women using the birth-control pill; a sexually transmitted disease 150

carrier a heterozygote who appears normal but who carries a recessive gene for a genetic disorder 25

carrying capacity the maximum number of organisms that an environment can support 184

cervical cap plastic cap inserted into the vagina to cover only the cervix for the purpose of preventing pregnancy 134

chancre a small papule with an open sore where an organism invaded the body 148

chancroid a sexually transmitted disease caused by a bacterium and characterized by ulcerated lymph nodes 149

chlamydia one of the most prevalent sexually transmitted diseases; caused by a bacterium and resulting in urinary and reproductive tract infection leading to PID 148

chorion the extraembryonic membrane that lies outside the amnion and becomes part of the placenta 114

chorionic villi testing examination and testing of embryonic blood from the chorionic villi 17

chromatids the two identical parts of a duplicated chromosome 5

chromosome rod-shaped body in the nucleus, particularly visible during cell division, that contains the genes 5

clitoris a slightly erectile organ of females that is homologous to the penis 108

codon a "triplet" of three nucleotides in messenger RNA that directs the placement of a particular amino acid into a polypeptide chain 48

coitus interruptus an abrupt withdrawal of the penis to prevent sperm from entering the vagina; a birth-control measure 135

color blind a genetically inherited disorder controlled by an X-linked allele, resulting in the inability to distinguish all colors visually 38

complementary base the nucleotide base that pairs with another. In DNA, adenine always pairs with thymine and guanine always pairs with cytosine, and vice versa. If RNA nucleotides pair with DNA nucleotides, adenine always pairs with uracil and guanine always pairs with cytosine, and vice versa 43

condom a sheath that fits over the erect penis, preventing the sperm from entering the vagina; a birth-control device 135

congenital defect any condition present at birth 114

congenital syphilis syphilis contracted during embryonic or fetal development 149

corpus albicans a body, white in color, that forms in the ovary from a corpus luteum that is no longer secreting progesterone 102

corpus luteum a body, yellow in color, that forms in the ovary from a follicle that has discharged its egg; it secretes progesterone 100

Cowper's glands glands that lie on either side of the urethra and secrete fluids that are a part of seminal fluid 93

Cri du Chat a syndrome in children caused by a shortened number-5 chromosome. Characterized by a small head, malformations of the head and body, mental defectiveness, and a cry similar to a kitten's meow 17

cryptorchindism a condition in which the testes remain in the abdominal cavity and do not descend into the scrotal sacs as is normal during embryonic development 87

cystic fibrosis a genetically inherited disorder characterized by abnormal mucous-secreting tissue 25

cytoplasm the fluid portion of the cell located outside the nucleus 5

cytosine (C) one of four nucleotide bases found in DNA and RNA; pairs complementarily with guanine (G) 43

D

desiccation drying out due to lack of water 163

diaphragm a cup-shaped birth-control device that fits over the cervix 134

diploid the 2N number of chromosomes; the complete or total number; twice the number of chromosomes found in gametes 8

DNA (deoxyribonucleic acid) the hereditary material 43

DNA probe a short section of single stranded DNA that is applied to cells or cell extracts in order to determine if complementary DNA is present 58

dominance hierarchy in behavior, the presence of a pecking order by which some animals exert control over other animals 174

dominant allele hereditary factor that expresses itself when the genotype is heterozygous 21

double helix a double spiral used to describe the three-dimensional shape of DNA 43

doubling time the number of years it takes for a population to double in size 182

Down syndrome a syndrome most often caused by the individual inheriting three number-21 chromosomes; characterized by an eyefold, a palm crease, and mental retardation 15

E

ectopic pregnancy an implantation of the embryo in a location other than the uterus 104

egg female gamete that contains a haploid number of chromosomes and is fertilized by male gamete, a sperm 8, 99

ejaculation the release of semen from an erect penis; transport of semen from the body 96

ejaculatory duct a duct that receives sperm from a vas deferens and secretions from a seminal vesicle and carries them to the urethra in males at the time of orgasm 92

embryo that stage of development from fertilization to the end of the second month   116

embryonic development human development from the time of fertilization to the end of the second month; the time when all major organs form   112

endocrine gland gland of internal secretion that produces hormones   71

environmental resistance environmental factors that prevent a population from achieving its biotic potential   184

enzyme protein molecule that speeds up a chemical reaction in cells   48

epididymis coiled tubule where sperm mature after being produced in the seminiferous tubules and before entering a vas deferens   90

episiotomy incision of the vaginal opening to facilitate birth   124

estrogen one type of female sex hormone, the other being progesterone   78

evolution genetic changes that occur in populations of organisms with the passage of time, resulting in increasing adaptation of the organism to the prevailing environment   163

extraembryonic membrane membrane that envelops the embryo and later the fetus   113, 169

feedback control a system of regulation by which the increase in a product leads to a decrease in its production and vice versa   78

fetal development development that occurs between the third and ninth months of pregnancy   112

fetus that stage of development from the end of the second month to birth   122

follicle ovarian structure than contains a germ cell in which the first stage of meiosis occurs in females   99

follicle-stimulating hormone (FSH) a hormone secreted by the anterior pituitary gland that stimulates the development of an ovarian follicle in a female or the production of sperm in a male   90, 103

foreskin skin covering the glans penis in uncircumcised males   93

gametogenesis production of gametes by the process of meiosis and maturation   8

gene a unit of heredity located on particular chromosomes and composed of DNA   5

genetic marker a difference in the sequence of DNA bases detected by restriction endonuclease enzymes that serves to indicate the inheritance of a genetic disorder because the marker and the disorder are almost always inherited together   63

genetic variation genotype changes such as those due to mutation and recombination   165

genital herpes a sexually transmitted disease caused by a herpes simplex virus and characterized by ulcers on the genitals   145

genital warts a sexually transmitted disease caused by papilloma virus and characterized by warts on the genitals   146

genotype the genetic makeup of an individual   21

germ cell a cell whose progeny undergo meiosis and maturation to become an egg in females or sperm in males   90

glans penis the terminal portion of the penis, covered by the foreskin in uncircumcised males; sexually sensitive and contains urethral opening through which either urine or semen exit the body   93

gonad an organ that produces sex cells; ovary and testis   78

gonadotropic hormone (GnH) hormone produced by the anterior pituitary that stimulates the gonads to produce gametes and hormones   78

gonadotropic-releasing hormone (GnRH) a hormone secreted by the hypothalamus that stimulates the anterior pituitary to release gonadotropic hormone (GnH)   78

gonorrhea a sexually transmitted disease caused by a bacterium that infects the reproductive tract; a cause of PID   147

Graafian follicle a mature ovarian follicle characterized by a fluid-filled cavity with the egg to one side   100

growth rate population growth that is calculated by subtracting the annual death rate from the annual birthrate   181

guanine (G) one of the four nucleotide bases in DNA and RNA that pairs complementarily with cytosine (C)   43

gumma a soft, gummy tumor occurring in tertiary syphilis   148

haploid the N number of chromosomes; half the diploid number; the number characteristic of gametes, which contain only one set of chromosomes   11

helper T-cells the type of lymphocyte that promotes the activity of other immune cells such as B lymphocytes, which produce antibodies   144

hemophilia an X-linked genetic disease characterized by excessive bleeding because the blood does not clot   38

hermaphroditic an animal having both male and female sex organs   166

herpes simplex virus the causative agent for genital herpes (type 2) and cold sores and fever blisters (type 1)   145

heterozygous genotype having two different alleles (e.g., *Aa*) for a given trait   21

homozygous dominant genotype having two dominant alleles (e.g., *AA*) for a given trait   21

homozygous recessive genotype having two recessive alleles (e.g., *aa*) for a given trait   21

hormone a chemical produced by one set of cells in the body that affects a different set of cells   71

human immunodeficiency virus (HIV) the causative agent for acquired immune deficiency syndrome (AIDS); causes a reduced number of helper T-cells in blood   144

hymen a thin, stretchable membrane partially covering the vaginal opening   108

hypothalamus a part of the brain that controls the pituitary gland and is involved in various emotional states   74

implantation an embryo having embedded itself in the uterine lining 106

impotency the inability to achieve or sustain an erection 96

in vitro fertilization (IVF) union of sperm and egg in laboratory glassware prior to implanting the developing embryo in the uterus of a female 141

infertility the inability to conceive and/or have as many children as one desires 140

innate inborn and inherited, as opposed to environmentally induced 171

interstitial cell stimulating hormone (ICSH) alternative name for LH in males; see luteinizing hormone (LH) 90

interstitial cells cells that lie between the seminiferous tubules in the testes and produce androgens 90

intrauterine device (IUD) a small coil or loop, usually made of plastic, that is placed in the uterus to prevent pregnancy 133

invertebrates animals without backbones such as flatworms, earthworms, and insects 166

karyotype a display of an individual's chromosomes arranged by pairs to show the number, size, and shape of the chromosomes 5

Klinefelter (XXY) syndrome a condition characterized by the sex chromosomes XXY, usually showing some feminization of features 19

L

labia majora two large folds of skin that constitute the outer lips of the vulva; homologous to scrotum of male 108

labia minora two small folds lying between the labia majora 108

larva an embryonic form that is capable of feeding and developing into an adult 166

Lesch-Nyhan syndrome an X-linked genetic condition characterized by self-mutilation 39

less developed country a country that is not yet industrialized and typically has a low standard of living and a high birth rate 184

luteinizing hormone (LH) the gonadotropic hormone, secreted by the anterior pituitary, that stimulates the corpus luteum in females and the interstitial cells in males 103

lymphocytes the type of white blood cell in the body that plays a major role in the development of immunity 143

meiosis cell division in which the four daughter cells have half the number and one of each kind of chromosomes as the original mother cell 8

menarche first occurrence of the menstrual discharge 108

menopause cessation of the uterine cycle 108

menstruation a monthly flow of blood caused by the periodic breakdown of the uterine lining in females 107

messenger RNA (mRNA) a ribonucleic acid complementary to genetic DNA and bearing a message to direct protein synthesis at the ribosomes 46

metafemale a female who inherits more than two X chromosomes 19

mitosis cell division in which the two daughter cells have the same number and kind of chromosomes as the mother cell 8

mons pubis a fatty prominence underlying the pubic hair 108

more developed country a country that is or is well on its way to being industrialized; usually characterized by a high standard of living and a low birth rate 184

motivation the drive to carry out a particular behavioral pattern 171

multiple alleles a genetic factor that occurs as more than two alternative forms at the same locus of a chromosome 30

muscular dystrophy a genetic disease characterized by a wasting away of the muscles 39

mutation a permanent gene change that is passed on to other generations of cells or organisms 26, 52

natural family planning a method of birth control based on abstinence during the fertile portion of a female's ovarian and uterine cycle 135

natural selection a process by which the environment favors the survival and reproduction of some members of a species as opposed to others 163

Neisseria gonorrheae the causative agent for a gonorrheal infection 147

neurofibromatosis (NF) a genetic disorder characterized by nonmalignant skin tumors 26

nondisjunction the failure of chromosomes to separate, especially during meiosis 17

nongonococcal urethritis (NGU) inflammation of the urinary tract which accompanies some disease other than gonorrhea 148

nonrenewable resource natural resource that is not periodically renewed and therefore is subject to depletion 186

Norplant a skin implant consisting of six silicone rubber tubes that leak progestin into the tissues to interrupt a woman's ovarian cycle; a form of birth control 133

nucleus a large organelle containing the chromosomes and acting as a control center for the cell 5

oncogenes genes that cause cancer usually because they code for a growth factor or some element in the cell that pertains to the reception or metabolic effect of a growth factor 53

oogenesis production of eggs in females by the process of meiosis and maturation 11

orgasm climactic sexual responses including both physical and emotional responses 97

ovarian cycle the follicular and hormonal changes that occur as the ovary responds to gonadotropic hormones during an approximately 28-day cycle 103

ovary female gonad; that portion of the pistil (female part of flower) that produces the ovules 99

oviduct tube that connects the ovaries and the uterus; this tube receives the egg following ovulation and is where fertilization occurs   104

ovulation the release of an egg from a Graafian follicle   100

oxytocin a hormone produced by the posterior pituitary that is necessary for milk letdown   74

P

parturition labor, followed by childbirth 123

pelvic inflammatory disease (PID) infection of the reproductive tract that has spread to the oviducts in females and the vas deferens in males; often results in sterility because of blocked tubes   134, 147

pelvis a bony cavity formed by the pelvic girdle along with the sacrum and coccyx 80

phenotype the appearance of an individual, which is a direct result of inherited genes   21

phenylketonuria a genetic disease characterized by the inability to metabolize phenylalanine, leading to mental retardation   26

pituitary gland an endocrine gland located at the base of the brain; sometimes called the master gland because it controls other endocrine glands   74

placenta the region where the embryo or fetus receives nutrients and discharges waste   114

plasmid extrachromosomal DNA that occurs in the form of a small ring in bacteria   55

polar body nonfunctioning cell that has little cytoplasm and is formed during oogenesis   13

polygenic inheritance a trait that is controlled by more than two genes   28

primary sex characteristics the sex organs of an individual   71

progesterone one type of female sex hormone, the other being estrogen   78

prolactin a hormone that stimulates the breast to produce milk; also called lactotropic hormone   125

prostate gland a gland that surrounds the urethra just below the bladder; secretes fluids that are a part of seminal fluid   93

protein a molecule made up of amino acids joined together in a specific sequence. Enzymes are usually proteins 48

puberty that time of life when the sex organs mature and the secondary sex characteristics appear   71

Punnett Square a grid that allows the determination of expected ratios of offspring by having all possible sperm fertilize all possible eggs   14

R

recessive allele allele that is not expressed unless it is inherited in double measure   21

recombinant DNA DNA that carries a new and different portion; particularly *E. coli* plasmid DNA carrying non-*E. coli* DNA   55

refractory period a period of time after ejaculation when the penis will not respond to sexual stimuli   97

releasing hormone hormone produced by the hypothalamus that controls the production of pituitary hormones   74

renewable resource a natural resource that is periodically renewed as water is renewed by the water cycle and food is renewed by agriculture   186

replication the process by which DNA is copied so that a daughter cell can receive a full complement of genes following cell division   45

ribosomal RNA (rRNA) the type of ribonucleic acid that is found in the ribosomes; therefore sometimes called structural RNA   46

ribosomes small particles in cells that are often associated with the endoplasmic reticulum and function to help bring about protein synthesis   43

RNA (ribonucleic acid) the type of nucleic acid that takes the genetic message to the ribosomes, is found in ribosomes, and transfers amino acids to the ribosomes, designated as mRNA, rRNA, and tRNA, respectively   43

scrotum a pouch of loose skin that contains the testes   87

secondary sex characteristics all those features that help distinguish males from females other than the sex organs themselves. For example, breasts, beard, musculature, and voice   71

seminal fluid semen; a fluid that contains sperm and secretions from the seminal vesicles, Cowper's glands, and prostate gland   92

seminal vesicle small gland that lies at the base of the bladder and secretes fluids that are a part of seminal fluid 92

seminiferous tubule long, thin tubule in the testes that produces sperm   87

Sertoli cells cells in interstitial cells of testes that protect and nourish sperm as they develop   90

severe combined immune deficiency syndrome (SCID) a genetically inherited disease characterized by the complete lack of immunity; the bubble-baby disease   39

sex chromosomes the chromosomes that determine sex (XX in females, XY in males)   5

sex hormones the hormones in females (estrogen and progesterone) and in males (androgens, i.e., testosterone) that account for the maturity of the primary sex characteristics and the maintenance of the secondary sex characteristics   71

sex-influenced trait trait that is more common to one sex than the other but that is not controlled by genes on the sex chromosomes; rather it is influenced by the presence or absence of sex hormones 40

sex linked genes or traits controlled by genes that are on the sex chromosomes 35

sexually transmitted diseases (STD) diseases caused by agents that are transmitted from one person to another during sexual intercourse or some other sexual activity   143

sexual reproduction reproduction that involves the union of gametes produced by two parents   8, 164

sickle-cell disease a genetic disorder characterized by sickle-shaped red blood cells   27

smegma a thick, cheesy secretion found under the foreskin of the glans penis   93

sociobiologist a biologist who applies evolutionary theory to animal and human behavior   177

sperm the male gamete that contains the haploid number of chromosomes and joins with the female gamete to start a new life   90

spermatogenesis production of sperm in males by the process of meiosis and maturation   11

sterilization a surgical procedure for birth control that renders the individual incapable of reproduction   129

survival of the fittest the likelihood of better-adapted organisms surviving and reproducing   163

syndrome a set of symptoms that characterizes a particular condition   15

syphilis a sexually transmitted disease caused by a bacterium and characterized by three stages with latent periods between; the last stage may have circulatory and nervous dysfunctions and large sores called gummas   148

Tay Sachs an inherited lysosomal storage disorder causing impairment of the nervous system   25

territoriality a type of behavior in which a member of a group stakes out a territory and prevents other members of the group from entering   172

tertiary sewage treatment plant a plant that treats human sewage even to the point that nutrient molecules are removed   187

testes the gonads of the male that develop in the abdominal cavity but descend into the scrotal sacs (scrotum)   87

testosterone the most potent of the androgens, the male sex hormones   78

thromboembolism presence of a blood clot in a circulatory vessel that prevents oxygen from reaching the heart muscle so that a heart attack occurs   133

thymine (T) one of four nucleotide bases in DNA that pairs complementarily with adenine (A)   43

transcription the formation of mRNA complementary to a DNA strand; when the code is transcribed into codons   48

transfer RNA (tRNA) a type of ribonucleic acid that transfers amino acids to the site of protein synthesis at the ribosomes   46

transgenic organisms organisms that have received a foreign gene so that they have a biochemical capability formerly lacking   60

translation the joining of amino acids in the order dictated by the DNA code and mRNA codons   48

Treponema pallidum the causative agent for syphilis   148

trichomoniasis a sexually transmitted disease caused by *Trichomonas vaginalis* characterized in females by a discharge and itching of the genitals   150

Trichomoniasis vaginalis protozoan (microorganism) that causes an infection of the vagina   150

triplet code a sequence of three DNA bases that stands for a particular amino acid so that a certain protein always has a certain sequence of amino acids   48

tubal ligation cutting of the oviducts, a method of sterilization in females   130

Turner syndrome a sex chromosomal abnormality in which a female has only one X chromosome (XO); characterized by short stature, web neck, and no breast development   17

umbilical cord a tubular structure that runs between the fetus and the placenta, containing the umbilical vein and umbilical artery   116

uterine aspiration the removal of the uterine contents by suction usually for the purpose of removing an implanted embryo   137

uterine cycle a monthly sequence of events that prepares the uterine lining to receive a developing embryo   106

uterus muscular organ, also called the womb, that lies between the bladder and rectum. Embryonic and fetal development occur in the uterus   106

vagina canal leading from the uterus to the vestibule that serves as the copulatory organ in females   108

vaginal sponge a type of birth-control device that consists of a sponge permeated by a spermacidal substance   134

vas deferens duct that lies between the epididymides and the ejaculatory duct where sperm are stored before entering the urethra during emission   90

vasectomy cutting of the vasa deferentia; a method of sterilization in males   129

vertebrates animals that have a backbone such as fishes, amphibians, reptiles, birds, and mammals   168

vestibule a cleft between the labia minora that contains the openings of the urethra and the vagina   108

vulva the region containing the external genitals of females   108

X linked a gene or a trait controlled by a gene located on the X chromosome   35

XYY male a male with the sex chromosomes XYY; characteristics include above-average height, persistent acne, and subnormal intelligence   19

Y linked a gene or a trait controlled by a gene located on the Y chromosome   35

yolk sac one of the extraembryonic membranes that in humans serves as the first location for the production of blood cells   114

Z

zero population growth no population growth, as when a population maintains the same number of births and deaths so that population size remains the same   185

zygote a cell resulting from fertilization that will develop into the new individual with the diploid number of chromosomes   8, 111

Credits

PHOTOGRAPHS

Table of Contents

Page vii: © Joe Sohn/The Image Works; **p. viii:** © Anne Rippy/The Image Bank; **p. ix:** Comstock, Inc./Georg Gerster

Part Openers

Part 1: © Joe Sohn/The Image Works; **Part 2:** © Anne Rippy/The Image Bank; **Part 3:** Comstock, Inc./Georg Gerster

Chapter 1

Opener: © Chuck Brown/Photo Researchers, Inc.; **1.7:** © Eric Grave/Photo Researchers, Inc.; **1.13a:** Jill Cannefax/EKM-Nepenthe; **1.15:** A. M. Winchester

Chapter 2

Opener: © Biophoto Associates/Photo Researchers, Inc.; **2.4a & b:** © Bob Coyle; **2.4c & d:** © Tom Ballard/EKM Nepenthe; **2.7a & b:** © Steve Uzzell III, 1982; **2.9a & b:** © Bill Longcore/Photo Researchers, Inc.; **2.10:** Editorial Enterprises; **2.11:** © Robert Burroughs; **2.12a & b:** © Stuart I. Fox

Chapter 3

Opener: © Manfred Kage/Peter Arnold, Inc.; **Box 3.1a:** © A. Robinson, National Jewish Hospital & Research Center; **Box 3.1b:** © Danny Brass/Science Source/Photo Researchers, Inc.; **3.4:** Baylor College of Medicine

Chapter 4

Opener: © C. Edelmann/LaVillette/Photo Researchers, Inc.; **4.10:** © Nancy Hamilton/Photo Researchers, Inc.

Chapter 5

Opener: © David Scharf/Peter Arnold, Inc.; **5.1a–d:** Courtesy of Genentech, Inc.; **5.3a:** © Sanofi Recherchere; **5.3b:** © Cetus Research Place; **5.3c:** © Phototake, Inc.; **5.5:** From C. Haudenschild, News and Comment, 1 fig., Vol. 246, pg. 747, Science, November 10, 1989, "Gore Tex Organoids and Genetic Drugs." Culliton, B. © 1989 by the AAAS

Chapter 6

Opener: © Petit Format/Nestle/Science Source/Photo Researchers, Inc.

Chapter 7

Opener: © Petit Format/Nestle/Science Source/Photo Researchers, Inc.; **7.2b:** © Biophoto Associates/Science Source/Photo Researchers, Inc.; **7.3:** © Dr. Gerald Schatten

Chapter 8

Opener: © Andy Walker, Midland Fertility Services/Science Photo Library/Photo Researchers, Inc.

Chapter 9

Opener: © Petit Format/Nestle/Science Source/Photo Researchers, Inc.; **9.5:** © John Moss/Photo Researchers, Inc.; **9.8a:** © Lennart Nilsson; **9.9:** Petit Format/Science Source/Photo Researchers, Inc.; **9.10:** © Petit Format/Nestle/Science Source/Photo Researchers, Inc.

Chapter 10

Opener: © Biophoto Associates/Science Source/Photo Researchers, Inc.; **10.4a:** © Ray Ellis/Photo Researchers, Inc.; **10.4b–f:** © Bob Coyle

Chapter 11

Opener: © Petit Format/Nestle/Science Source/Photo Researchers, Inc.; **11.2:** © Charles Lightdale/Science Source/Photo Researchers, Inc.; **11.3–11.5:** Health, Education, Welfare, Center for Disease Control, Atlanta, GA; **11.6:** Reproduced by permission from Donaldson, David, Atlas of External Diseases of Eye, Vol. 1 Congenital Anomalier & Systematic Diseases, St. Louis, 1966, The C. V. Mosby Co.; **11.7a–c:** © Carroll H. Weiss/Camera M.D. Studios

Chapter 12

Opener: © Petit Format/Nestle/Science Source/Photo Researchers, Inc.; **12.6:** © Roger K. Burnard/Biological Photo Service; **12.9:** © Mitch Reardon/Science Source/Photo Researchers, Inc.

Chapter 13

Opener: © Petit Format/Science Source/Photo Researchers, Inc.; **13.3:** © Tom McHugh/Photo Researchers, Inc.; **13.7:** © Mitch Reardon/Photo Researchers, Inc.; **13.8:** © Spencer Grant/Photo Researchers, Inc.

Chapter 14

Opener: © Suzanne Szasz/Photo Researchers, Inc.; **14.9:** © Harold V. Green/Valan Photos

ILLUSTRATIONS

Chapter 1

1.14: From Robert F. Weaver and Philip Hedrick, *Genetics.* Copyright © 1989 Wm. C. Brown Publishers, Dubuque, Iowa. All Rights Reserved. Reprinted by permission.

Chapter 3

3.5: From E. Peter Volpe, *Biology and Human Concerns,* 3d edition. Copyright © 1983 Wm. C. Brown Publishers, Dubuque, Iowa. All Rights Reserved. Reprinted by permission.

Chapter 6

6.5: From Kent M. Van De Graaff and Stuart Ira Fox, *Concepts of Human Anatomy and Physiology,* 2d edition. Copyright © 1989 Wm. C. Brown Publishers, Dubuque, Iowa. All Rights Reserved. Reprinted by Permission. **6.11:** From Kent M. Van De Graaff and Stuart Ira Fox, *Concepts of Human Anatomy and Physiology,* 2d edition. Copyright © 1989 Wm. C. Brown Publishers, Dubuque, Iowa. All Rights Reserved. Reprinted by permission. **6.13:** From Kent M. Van De Graaff and Stuart Ira Fox, *Concepts of Human Anatomy and Physiology,* 2d edition. Copyright © 1989 Wm. C. Brown Publishers, Dubuque, Iowa. All Rights Reserved. Reprinted by permission.

Chapter 7

7.1: From John W. Hole, Jr., *Human Anatomy and Physiology,* 5th edition. Copyright © 1990 Wm. C. Brown Publishers, Dubuque, Iowa. All Rights Reserved. Reprinted by permission. **7.2c:** From Kent M. Van De Graaff and Stuart Ira Fox, *Concepts of Human Anatomy and Physiology,* 2d edition. Copyright © 1989 Wm. C. Brown Publishers, Dubuque, Iowa. All Rights Reserved. Reprinted by permission. **7.2d:** From John W. Hole, Jr., *Human Anatomy and Physiology,* 5th edition. Copyright © 1990 Wm. C. Brown Publishers, Dubuque, Iowa. All Rights Reserved. Reprinted by permission.

Chapter 8

8.2a: From Kent M. Van De Graaff and Stuart Ira Fox, *Concepts of Human Anatomy and Physiology,* 2d edition. Copyright © 1989 Wm. C. Brown Publishers, Dubuque, Iowa. All Rights Reserved. Reprinted by permission.

Chapter 9

9.4: From John W. Hole, Jr., *Human Anatomy and Physiology,* 5th edition. Copyright © 1990 Wm. C. Brown Publishers, Dubuque, Iowa. All Rights Reserved. Reprinted by permission. **9.11:** Original Artwork Courtesy Carnation Company, Los Angeles, California. Copyright © Carnation Co., 1962. **9.12:** Original Artwork Courtesy Carnation Company, Los Angeles, California. Copyright © Carnation Co., 1962. **9.13:** Original Artwork Courtesy Carnation Company, Los Angeles, California. Copyright © Carnation Co., 1962.

ILLUSTRATORS

Chris Creek

2.1.

Fineline Illustrations, Inc.

3.3; 10.1.

Anne Greene

1.11; 7.8; 9.6, 9.7.

Kathleen Hagelston

1.5, p. 20; 2.12a–b, 2.13a–b; 3.1; 4.2, 4.3, 4.4, 4.5, 4.7, 4.8, 4.9; 6.1a–b, 6.6, 6.8a–c, 6.9, 6.10, 6.12; 7.4, 7.5, 7.6, 7.7a–b, Box 7.1; 8.3, 8.6, Box 8.1; 9.1, 9.3, 9.14, 9.15, 9.16, 9.17; A.1, A.2, A.3, A.4, A.5, A.6; 10.2, 10.3, 10.5, 10.6a–b, 10.7a–b, 10.8; 12.2, 12.7, 12.8; 14.2, 14.6.

Hagelston–O'Keefe

4.1; 5.2, 5.6.

Carlyn Iverson

1.2; 6.7; 7.9; 8.2b, 8.7, 8.8; 9.8b; Box 10.1; 12.3, 12.5; 13.2, 13.4a–e, 13.6.

Iverson–Rolin

1.4, 1.6a–f, 1.8, 1.9a–b, 1.10.

Ruth Krabach

9.4.

Marjorie Leggitt

1.13b.

Steve Moon

1.1; 6.13.

Precision Graphics

2.8; p. 33; 4.6; 5.7a–b; 13.1; 14.7.

Rolin Graphics

p. 8, 1.12, 1.14; 2.5, 2.6, 2.14a–d; 5.4; 6.4; 7.2a, 7.2c–d; 8.5; 9.2, Box 9.1; 11.1; 12.1; 13.5; 14.3, 14.5, 14.8.

Mike Schenk

6.11a–b.

Tom Waldrop

6.3, 6.5, Box 6.1; 7.1; 8.1, 8.2a.

Index

C

Cancer, 53, 95, 137, 146, 147
Candida albicans, 150
Capacitation, 111
Carriers, 25, 36
    HIV, 144, 155–56
Carrying capacity, 182–84
Cell
    chemical reaction in, 48
    cycle, 9
    division, 8–14
    human, 6
    human-mouse, 63
Centromere, 5, 9
Cervical cap, 129, 134
Cervix, 106
    cancer, 146, 147
    dilation, 123–24
    effacement, 124
Chancre, 148, 149
Chancroid, 149–50
Chlamydia, 148
Chlamydia trachomatis, 148
Chlamydomonas, 165
Chloasma, 133
Chorion, 112, 114, 116, 119,
    120, 169
Chorionic villi, 17, 26, 119
Chromatids, 5, 8–9, 11, 17
Chromosome, 5
    abnormalities, 15–19
    analysis, 119
    autosomal, 5, 14, 21
    chromatid, 5, 8–9
    deletion, 17
    diploid number, 8
    duplication, 9
    in egg and sperm, 4, 5, 8,
        17, 26, 90
    haploid number, 11, 13
    mapping, 59, 63–64
    in meiosis, 11–14
    in mitosis, 8–14
    nondisjunction, 17
    in oocyte, 7, 17, 25, 26, 99
    sex, 5, 14, 17–19
    in spermatocyte, 11, 90
    structure, 15, 17, 21, 44, 63
    X, 5, 11, 13, 14, 17, 30, 35,
        36, 37, 71
    XO, 17
    XXX, 17, 19
    XXY, 17, 19
    XYY, 17, 19
    Y, 5, 11, 14, 17, 35, 71
    in zygote, 8

Circulation, fetal, 114–16, 118,
    124
Circumcision, 93, 95
Cleft palate, 30
Climateric, 90
Clitellum, 167
Clitoris, 108, 109
Clomid, 102
Cloning, 9, 55, 57
Clotting factor VIII, 58
Codons, 48, 50
Coitus interruptus, 129, 135
Cold sores, 145
Color blindness, 38
Colostrum, 123, 125
Complementary base, 43
Conception, 112
Condom, 129, 135
    female, 137
    to prevent spread of disease,
        145, 148, 149, 155,
        159
Congenital defects, 114
Congenital syphilis, 149
Contraception, 129. *See also*
        Birth control
Contragestation, 137
Copulation, 166
Corona radiata, 100, 111
Corpora cavernosa, 93
Corpus albicans, 102
Corpus luteum, 100, 103, 104,
    106, 107, 108
    HCG and, 112
Corpus spongiosum, 94
Cortex, 99
Cortisol, 74
Cortisone, 123
Country, less developed and
        more developed, 184,
        185
Cowper's gland, 88, 92, 93, 96
Crabs, 150
Cream, spermicidal, 129, 135
Cri du Chat, 17, 18
Cryptorchidism, 87
Cystic fibrosis, 25
Cytoplasm, 5, 13, 14, 43, 46,
    50, 111
Cytosine (C), 43, 45, 46

D

Dalkon Shield, 134
Danazol, 138
Daughter cell, 8, 9, 11, 13

DDI, 145, 158
Defect
    birth, 114, 119
    congenital, 114
    genetic, 119
Deletion, 17
Deoxyribose, 43
DES, 137
Desiccation, 163
Development, 112–23
    embryonic, 112–13, 116–21
    fetal, 112–13, 122
Diaphragm, 129, 134
Dideoxyinosine, 158
Dilation, 140
Diplococcus, 147
Diploid number, 8
Disorder, genetic
    autosomal dominant, 26–27
    autosomal recessive, 24–26
    detecting with DNA probes,
        59
    genetic marker and, 63–64
    metabolic and structural,
        52–53
    mother's age and, 99
    polygenic, 30
    sex-linked, 36–39
    simple autosomal, 24–27
    treating with biotechnology,
        60–62
    X-linked recessive, 36–40
DMD. *See* Duchenne muscular
        dystrophy
DNA (deoxyribonucleic acid),
    43
    fingerprinting, 59
    functions, 45–46
    probe, 58–60, 64
    protein synthesis, 46, 48–52
    recombinant, 55, 143
    replication, 45–47
    structure of, 43–45
    triplet code, 48
Dominance, 172–74
    autosomal, 24, 26–27
    heterozygous, 21, 24
    hierarchy, 174
    homozygous, 21, 22
    incomplete, 27–28
Dominant allele, 21
Double helix, 43, 46
Down syndrome, 15, 16, 17
Drug
    impact on fetus, 114, 116
    postcoital, 137
    treatment of AIDS, 158
Duchenne muscular dystrophy
    (DMD), 37, 38, 39,
    64

Duck-billed platypus, 169
Ductus arteriosus, 116
Dystrophin, 39

E

Earlobe, 21, 22–24
Earthworm, 167
Effacement, 124
Egg, 8, 13, 14, 99, 100
    fertilized, 111
    invertebrate, 166, 167
    mammal, 177
    passage into oviduct, 104–5
    screening, 119
    vertebrate, 168, 169
Ejaculation, 14, 96–97, 111
Ejaculatory duct, 88, 90, 92
Elephant man disease, 26
Embolus, 133
Embryo, 113, 116, 177
    early months, 120–21
Emission, 96
Endocrine gland, 71–78
Endometriosis, 138–39, 140
Endometrium, 106, 107, 108,
    138
Endoplasmic reticulum, 48
Endothelial cell, 62
Environmental resistance, 184
Enzyme, 48, 52
    acrosome, 111
    DNA ligase, 55
    restriction endonuclease, 55,
        63
Epididymis, 88, 90
Episiotomy, 124
Erection, 94
Erythropoietin, 58
Estrogen, 78, 83, 99, 100, 104
    in birth-control pill, 130,
        133
    endometriosis and, 138
    low levels of, 108
    placenta and, 112
    uterine cycle and, 106, 107
Estrus, 178
Evacuation, 140
Evolution, 163
    of sexual reproduction,
        163–70
    survival of fittest in, 163
Excitotoxin, 27
Expulsion, 96
Eye infection, 146, 147–48